PHYSICS LECTURES

Concise Outlines for College & University

Robert W Finkel

St. John's University, NY

ISBN-13: 978-1466218277
ISBN-10: 1466218274

Chaion Analytics

Printed in the United States of America.

Betty and Abe

In Loving Memory

Physics Lectures

Preface

Modern textbooks for college or university physics are huge and densely packed with equations and problems. These books typically outweigh world history or literature texts.

Publishers seem convinced that more is better. More artistic figures, more advanced applications, more topics, more heavy magazine paper, more color, more specialized equations, more problems, more pages, and—oh yes—more cost.

Faced with this excess, students usually rely on lectures to guide them through the simple fundamentals of the subject and standard model problems.

Physics Lectures derives from a set of lecture notes presenting basic theory and typical problems. It is intended as a framework for the course. Little space is given here to development and background, the live flesh of the subject that covers this skeleton outline.

I kept the format and flavor of the lecture notes in this book. The result is the antithesis of encyclopedic texts that are written to appease specialists and enrich publishers.

Physics Lectures

Physics Lectures is a bare-bone, accessible, and inexpensive alternative. It enjoys outstanding classroom successes.

The fundmentals for calculus and non-calculus treatments overlap so that the differences are largely in notation. Some topics or problems are the exclusive province of a calculus based course and these sections are clearly identified. They are easily disregarded for non-calculus sections.

Exercises are designed to be part of the pedagogical fabric of the book and much of my lecture time was spent reviewing these in class. Most of the exercises have answers attached or have solutions written out in full.

Standard texts have copious end-of-chapter problems from which instructors select assignments. I included small sets of unsolved problems at the chapter ends to accommodate this practice. This should make this book self-contained for instructors who will use it as their primary text.

Note to instructors: I recognize that in adopting this book you put your students' learning above your own advanced interests. I applaud your dedication and wish you success.

Robert W Finkel
New York

Table of Contents

1

MOTION ALONG A LINE

This chapter introduces the mathematical description of a moving object. The physical causes of such motions are left to later chapters.

Overview

When a particle is traveling along a line, the motion is completely described by specifying its position x at any time t,

$$x = x(t) .$$

Note that this notation does *not* mean "x times t." Rather, it says "x is a function of t." For example, the equation $x = 3t^2$ describes an object that at time $t=1$ is at position 3 and at $t=2$ is at position 12, etc.

Starting from $x(t)$ we will be able to calculate the velocity v at time t written as $v = v(t)$ and the "acceleration" or rate of change of velocity a,

$$a = a(t) .$$

We are often called upon to reverse the process; given $a(t)$, find $x(t)$. This happens because a law of motion ($F=ma$) lets us calculate acceleration from a known force F and mass m. The request to "find the motion" then means calculate position and velocity from acceleration (and other initial conditions).

Average & Instantaneous Velocity

Average velocity is the change in displacement divided by the change in time. In equation form,

$$\bar{v} = \frac{\Delta x}{\Delta t}$$

The Δ (delta) notation does not indicate multiplication by Δ. Instead, it should be read as "a change in" the quantity that follows.

1.1   The equation $x = 4.9\,t^2$ gives the distance in meters that a stone falls in t seconds. Find the average velocity of the stone between 2 and 2.01 seconds. (Use all significant figures and include units in the answer.)

Instantaneous velocity is is the velocity of an object at a particular point or a particular time (like the reading of a speedometer). When the term *velocity* is used without further specification, we usually mean instantaneous velocity defined in calculus terms by the expression.

Calculus students will recognize the formal expression for instantaneous velocity,

$$v = \operatorname*{Limit}_{t \to 0} \frac{\Delta x}{\Delta t} \equiv \frac{dx}{dt}$$

1.2   Repeat the calculation of problem 1.1 except use the symbol Δt instead of the numerical time 0.01 sec. Simplify the algebra and then set $\Delta t = 0$ without dividing by zero. You will have computed the instantaneous velocity of the stone at t=2 sec. Calculus students should do this also by a standard formula. *

---

* Recall that $\dfrac{d}{dt} t^n = nt^{n-1}$

1 MOTION ALONG A LINE

Acceleration

Acceleration is the rate of change of velocity, $a = \Delta v / \Delta t$, or in calculus notation,

$$a = \frac{dv}{dt}$$

The MKS units for acceleration are m/s² (meters per second squared).

1.3   Calculus based problem: Compute the acceleration of the stone in problem 1.1 at t=2 sec and t=3 sec. (Include units in answer.)

Constant Acceleration Equations

Most often (but not always), we will be working with constant accelerations. Almost any motion problem involving constant acceleration can be solved using one or more of the following four equations where the symbols x_0 and v_0 represent the initial (beginning) values of position and velocity:

$$x = \frac{v_o + v}{2} t$$
$$x = x_o + v_o t + \tfrac{1}{2} a t^2$$
$$v = v_o + at$$
$$v^2 = v_o^2 + 2a(x - x_o)$$

It is a great help in solving these problems to write the *symbols* for each quantity. This will help you choose the equation(s) that are appropriate to the problem. Watch for catchwords in problem statements like "from rest" $(v_o = 0)$ and "uniform acceleration" (constant a).

1.4 An acorn falls from rest with acceleration 9.8 m/s² and strikes the ground at 19.6 m/s. For how many seconds was the acorn falling?

Often the initial position of an object is not specified. You can usually choose $x_o = 0$.

1.5 A truck moving at 10 m/s applies the brakes and stops after skidding 13.5 m. Find the uniform deceleration. (Include units in answer.)

1.6 An acorn falls from rest with acceleration 9.8 m/s² and strikes the ground at 9.8 m/s. How far did it fall?

1.7 An acorn falls from rest with acceleration 9.8 m/s² and strikes the ground in 1.0 sec. How far did it fall?

Free Fall

All objects that fall near the surface of the earth have the same downward acceleration "g" (this assumes approximately no air resistance):

$$g = 9.8 \text{ m/s}^2$$

Even objects that are thrown upward have a *downward* acceleration of 9.8 m/s². Free-fall is a special case of constant acceleration. Usually, but not always, we prefer to take "up" as positive so

$a = -g$ (when "up" is positive in free-fall problems)

The highest point in a free-fall is the point at .which the particle stops moving up; that is,

$v = 0$ (highest point)

1 MOTION ALONG A LINE

1.8   A champagne cork shoots straight up out of its bottle at 8.85 m/s. Find
the time it takes to reach its highest point,
the height it attains above its point of release,
the velocity with which it returns to its starting point,
the time it takes from its release to its return.

Answers

1.1   19.649 m/s
1.2   19.6 m/s
1.3   9.8 m/s^2 , 9.8 m/s^2
1.4   2 s
1.5   -3.7 m/s^2
1.6   4.9 m
1.7   4.9 m
1.8   0.903 s, 4.0 m, -8.85 m/s, 18.06 s

Problems

1.   If a body starts from rest with a uniform acceleration of 2 m/s^2, find its velocity at the end of 3 minutes.

2.   A body starts with a velocity of 300 m/s. If it comes to rest in 1 minute 2.5 seconds, find the uniform negative acceleration.

3.   A body has an initial velocity of 8 m/s; find its velocity at the end of 1, 2, 3, and 8 seconds respectively, if a equals 9.8 m/s^2.

4.   A body with uniform acceleration acquires a velocity of 10 m/s in moving a distance of 25 m from rest, find the acceleration.

5.   In what time will a body moving with a uniform acceleration of 9 m/s^2 traverse 250 meters?

6.   A cannon ball has s muzzle velocity of 400 m/s; the length of the gun traversed by the ball is 3 m. Find (a) the acceleration on the assumption that it is uniform; (b) the time of traversing the gun.

7.   A particle has a uniform acceleration of 20 cm per second per second, and an initial velocity of 30 cm/s. Find (a) the velocity after 16 seconds; (b) the time required to travel 300 cm; (c) the change in velocity in traversing that distance.

8.   A body is projected upward with any velocity and t and t' denote times during which it is respectively above and below the middle point of its path; find the value of t/t'.

9.   A body dropped from the top of a tower 20 m high reached the bottom of a well within the tower in 3 s; find the depth of the well.

10.   A body is projected vertically upward with an initial velocity of 250 m/s. Find
      (a) how high it will rise;
      (b) the.time of ascent;
      (c) when it will be 350 m above the starting point.

11.   A body thrown vertically upward passes a point 10 m from the starting point with a velocity of 20 m/s. Find (a) how much farther it will go; (b) its initial velocity.

2

VECTORS

This chapter is an interlude on our way to describing motion in more than one dimension. For instance, vectors will enable us to reduce two-dimensional motions like the trajectory of a baseball to two simple one-dimensional motions.

Vectors & Scalars

For our purposes, a *vector* is a quantity having (i) a magnitude or size and (ii) a direction. Example: *velocity* is a vector with magnitude called *speed* (say 65 km/hr) and a direction (say 20° East of North). Other vectors include displacement and force. In contrast, quantities that are fully described by magnitude alone are called *scalar* quantities. Examples: volume, time, density.

2.1 Identify the following as scalar or vector quantities:
(a) mass, (b) temperature, (c) weight, (d) acceleration

Vector Representation

Vectors are represented with arrows like ⟶ and ╱. The length of the arrow represents the magnitude and the orientation of the arrow represents the direction of the vector.

It is an important property of vectors that they may be moved around in space (or on paper) providing that the length and orientation are not changed.

2.2　　Exercise: Move the vectors so that the base of **B** is at the tip of A and the base of **C** is at the tip of **B**.

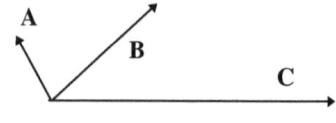

Vector Addition: Graphical Approach

To add vector **A** to vector **B**, put the base of **B** at the head of **A**; the sum **A+B** is a vector joining the base of **A** to the head of **B**.

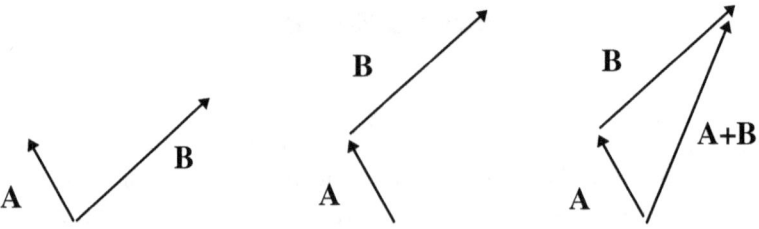

The vector (**A+B**) that represents a sum of vectors is called the *resultant* of the original vectors.

2.3　　Sketch the resultant of the vectors shown:

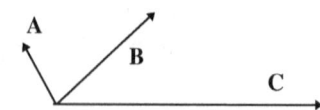

2 VECTORS

Vector Components

We have been concerned with the problem of constructing a resultant from two given vectors. Now we turn our attention to the problem in reverse; given a vector, find two "component" vectors whose resultant is the original vector. In particular, we want the two component vectors to be perpendicular to each other. The diagram shows the x-and y-components of vector **A**.

Recall that:

$$\sin\theta = \frac{\text{opp}}{\text{hyp}}$$

$$\cos\theta = \frac{\text{adj}}{\text{hyp}}$$

$$\tan\theta = \frac{\text{opp}}{\text{adj}}$$

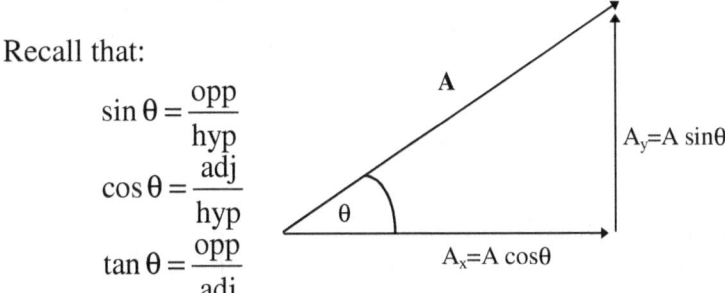

2.4   Compute the components of a force of 5 N directed at an angle of 53° above the horizontal toward the right.

Vector Addition by Components

It is easy to add several x-component (or y-component) vectors together because only ordinary additions and subtractions are needed.

2.5   Exercise: Find the resultant of two x-components:
a) 12 units to right and b) 3 units to right.

2.6   Exercise: Find the resultant of two y-components:
a) 5 units up and b) 4 units down.

This suggests the component method for adding vectors.

(1) Find the components of the vectors to be added and replace the originals with these components.

(2) Add the x-and y-components. Only one of each component will remain regardless of how many vectors were summed.

(3) Find the resultant from the x-and y-components. This requires that you find the length of the resultant and its angle of orientation.

2.7   A woman walks 5 m at $37°$ East of South and then 13 m at $22.6°$ North of East. Find her resultant displacement.

Unit Vectors

Unit vectors are used to describe vectors in terms of components. A unit vector in the x-direction is usually written **i** and has a length "1." Similarly, unit vectors in the y-and z-directions are denoted by **j** and **k**. For example, we can describe a vector **V** with an x-component of 3 and a y-component of 4 as follows:

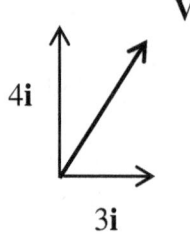

$$V = 3i + 4j$$

Vectors are particularly easy to add in this form. Let $\mathbf{A} = A_x\mathbf{i} + A_y\mathbf{j}$ and $\mathbf{B} = B_x\mathbf{i} + B_y\mathbf{j}$, then

$$\mathbf{A} + \mathbf{B} = \left(A_x + B_x\right)\mathbf{i} + \left(A_y + B_y\right)\mathbf{j}.$$

2. 8   Given $\mathbf{V} = 3\mathbf{i} + 4\mathbf{j}$ and $\mathbf{U} = -3\mathbf{i} + 6\mathbf{j}$, find
         (a) $\mathbf{V} + \mathbf{U}$, (b) $\mathbf{V} - \mathbf{U}$

2 VECTORS

Dot Product

The dot product (or, more generally, the scalar product) is a vector operation we often need in order find the projection of one vector on another or the angle between two vectors. The definition is,

$$\mathbf{A} \bullet \mathbf{B} = |\mathbf{A}||\mathbf{B}|\cos\theta,$$

where the vertical bars indicate vector magnitudes and θ is the angle between the two vectors.

2. 9   Use the definition of the dot product to show the following:

$$\mathbf{i} \bullet \mathbf{i} = \mathbf{j} \bullet \mathbf{j} = \mathbf{k} \bullet \mathbf{k} = 1$$
$$\mathbf{i} \bullet \mathbf{j} = \mathbf{i} \bullet \mathbf{k} = \mathbf{j} \bullet \mathbf{k} = 0$$

2. 10   Given $\mathbf{V} = 3\mathbf{i} + 4\mathbf{j}$ and $\mathbf{U} = -3\mathbf{i} + 6\mathbf{j}$, find $\mathbf{U} \bullet \mathbf{V}$.

2. 11   Use the fact that $\mathbf{A} \bullet \mathbf{A} = |\mathbf{A}|^2$ to find the length (magnitude) of $\mathbf{V} = 3\mathbf{i} + 4\mathbf{j}$.

2. 12   Given $\mathbf{V} = 3\mathbf{i} + 4\mathbf{j}$ and $\mathbf{U} = 3\mathbf{i} - 4\mathbf{j}$, find the angle between $\mathbf{U}$ and $\mathbf{V}$.

Answers

2.1   scalar, scalar, vector! (force directed down), vector
2.3   ↗
2.4   x-component 3, y-component 4
2.5   15 right
2.6   1 up
2.7   15.03 m at $3.8°$ North of East
2.8 $10\mathbf{j}$, $6\mathbf{i} - 2\mathbf{j}$
2.10   15
2.11   5
2.12 $101.5°$

Problems

1.   A vector drawn in an Easterly direction has a length of 20 cm, and one drawn Northeast has a length of 30 cm, What is their vector sum?

2.   Given vectors $\mathbf{A} = \mathbf{i} + 2\mathbf{j}$ and $\mathbf{B} = 2\mathbf{i} + \mathbf{j}$: (a) determine the angle between them, (b) find their lengths, (c) find their resultant and (d) determine the angle of the resultant with the x-axis (vector $\mathbf{i}$).

MOTION IN A PLANE

The equations that describe motion in 2 or 3 dimensions are simple generalizations of those that describe motion along a line.

Constant Acceleration in 2-Dimensions

Three sets of important constant acceleration equations in 2-D.

$$v_x = v_{ox} + a_x t$$
$$v_y = v_{oy} + a_y t$$

$$x = x_o + v_{ox} t + \tfrac{1}{2} a_x t^2$$
$$y = y_o + v_{oy} t + \tfrac{1}{2} a_y t^2$$

$$(v_x)^2 = (v_{ox})^2 + 2a_x(x\text{-}x_o)$$
$$(v_y)^2 = (v_{oy})^2 + 2a_y(y\text{-}y_o)$$

These component equations are identical to the constant acceleration expressions for motion along a straight line. You can view them as two separate motions along straight lines-one for the *x*-direction and one for the *y*-direction.

Projectile Motion

Ideal projectiles have no acceleration in the horizontal direction while having the same acceleration as free-fall in the vertical direction:

$$a_x = 0, \quad \text{and} \quad a_y = -g \quad \text{taking "up" as positive}$$

3.1   A cat jumps from a ledge in a horizontal direction at 4 m/s and strikes the ground 1.0 sec later. Find (a) the horizontal distance traveled by the cat and (b) the height of the ledge.

3.2   A cat jumps from a ledge in a horizontal direction at 4 m/s and strikes the ground at a horizontal distance of 4 m from the edge. Find (a) the velocity with which the cat strikes the ground and (b) the height of the ledge.

3.3   A golf ball is driven from the ground with a velocity of 29.4 m/s at an angle of $60°$ with the horizontal direction. (a) Find the time it takes to strike the ground, (b) Find the horizontal distance traveled by the ball, and (c) the highest point attained.

3.4   A golf ball is driven from the edge of a rooftop with a velocity of 29.4 m/s at an angle of $60°$ above the horizontal direction. It lands 103 m horizontally from the impact point. Find (a) the time it takes to strike the ground and (b) the height of the building

3.1   4 m, 4.9 m
3.2   10.6 m/s at $68°$ below the horizontal, 4.9 m
3.3   5.2 s, 76 m, 33 m
3.4   7 s, 62 m

Problems

1.   The muzzle of a gun on a ship's deck is 8.06 m above the water. A shell fired horizontally strikes the water 300 m away.What is its muzzle velocity?

2.   A plane traveling parallel to the ground at 260 m/s drops a package that reaches the ground in 45 seconds. Assume neglegble air resistance and
(a) find the altitude of the plane.
(b) find the horizontal distance traveled by the package.

3.   A golf ball is projected with a velocity of 33 m/s from a height 22 m above the ground at an angle of elevation of $30°$ with the horizontal. Find when and where it will strike the ground.

PHYSICS LECTURES

NEWTON'S LAWS

Newton's three laws of motion enable us to predict the future motion of an object or a collection of objects. In order to apply the laws in general, we need to specify the (a) forces on the objects, (b) masses of the objects, and (c) initial conditions of the objects (that is, their positions and velocities at the instant we wish to call the initial time).

First Law of Motion

The First Law describes motion when there are no outside forces on the object.

1st Law of Motion
A moving object will continue to move with constant velocity (that is, in a straight line and unchanging speed) until it is acted upon by an outside force. An object at rest will remain at rest until it is acted upon by an outside force.

4.1.   A force-free meteor moves directly North at 100 m/s at 12 PM. What is its velocity at 2 PM?

Second Law of Motion

Force is a measure of the push or pull exerted on an object. Force is a vector because it has a direction (the direction of the push or pull) and a magnitude or size (this can be measured in pounds). For example, the force due to gravity on a 15 lb dog is called *weight*—its direction is straight down and its magnitude is 15 lb. The unit of force in standard international units (SI or MKS units) is the *Newton*, N.

The Second Law says that acceleration of an object is proportional to the force applied to it [$F \propto a$ or $F = ma$]. The proportionality constant, m, is called the *mass* and it has a different value for different objects.

Definition: *Mass* is a measure of the resistance of an object to being moved (sometimes called *inertia*). Mass is closely related to weight, but mass does not have a direction and does not change when the object is moved from the earth to the moon (weight does). The units of mass are *kilograms*, kg.

2nd Law of Motion
$$F = ma$$
Force is proportional to acceleration and the proportionality constant is the mass of the object.

4.2.   A force of 10 N is applied to a 2 kg mass.
(a) Calculate the acceleration of the mass.
(b) If the mass starts from rest, how far does it travel in 3 seconds and how fast is it going?

4.3.   Derivation of $w = mg$: The force on an object near the earth's surface due to gravity is the object's weight; w. The acceleration of the object has magnitude g when no other forces are present. Write a relation between w, g, and the mass of the object.

The last problem shows that weight w and mass m are related by
$$w = mg$$
Weight is distinguished from mass; weight changes when the object is on another planet or is far from the earth, but the mass is taken to remain the same everywhere.

4.4.   What is the weight of a 50 kg barbell?

The force in $\mathbf{F} = m\mathbf{a}$ represents the sum of all forces acting on an object. This is often written explicitly as
$$\Sigma \mathbf{F} = m\mathbf{a}$$
Where the summation sign indicates that we must take the vector sum of all forces on the mass.

4.5.   The diagram is an overhead view of a sled. Two dogs pull the 30 kg sled by ropes. King pulls at $60°$ from the x-axis with a force of 20 N. Quantum

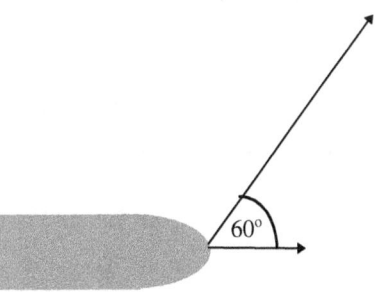

pulls along the x-axis with a force of 5 N. Find the acceleration of the sled (assume the motion is frictionless).

4.6. A 35 kg mass is pushed along a frictionless surface as shown. Force F in the diagram is 40 N. Find acceleration a.

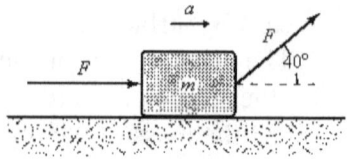

Third Law of Motion

3rd Law of Motion

When object A exerts a force on object B, then object B exerts a force on object A which is equal in magnitude, but opposite in direction. Note: The Third Law is often referred to as the law of action and reaction.

Each force in the pair is referred to as the *reaction force* of the other. For example, when a hand is pressed against a wall, it exerts a force of, say, 5 LB on the wall. The wall then exerts a 5 LB force on the hand, but in the opposite direction.

4.7. A coconut falls toward the ground. What are the equal and opposite forces involved? Hint: what large object exerts a force on the coconut?

Exercise: Explain rocket propulsion in terms of the 3rd Law.

4.8.   Consider a buggy being pulled by a horse. Which is correct?
(a) The horse can pull the buggy forward only if the horse weighs more than the buggy.
(b) The horse pulls forward slightly harder than the buggy pulls backward on the horse, so they move forward.
(c) The horse pulls before the buggy has time to react so they move forward.
(d) The force on the buggy is as strong as the force on the horse. The horse is joined to the Earth by its hoofs, while the buggy is free to roll on wheels.

Answers

4.1	100 m/s North
4.2	5 m/s^2, 22.5 m, 15 m/s
4.3	$w = mg$
4.4	490 N
4.5	0.76 m/s^2 directed 49° above x-axis
4.6	2.0 m/s^2
4.7	the coconut pulls the Earth "up"
4.8	d

Problems

1.   What is the acceleration when a force of 3.6×10^{-4} N acts on a mass of 4 gms. ? How far will the mass move in 10 seconds?

2.   A force of 6.0×10^{-4} N acts on a body for one minute and imparts to it a velocity of 9 m/s. What is the mass of the body?

3.   A mass of 650 gm is attached to a spring balance that is carried up at such a rate that the balance indicates 750 gms. What is the acceleration upward?

4   An elevator starts to descend with an acceleration of 3 meters per second per second. Find the force on its floor of a 75 kg man. What would be his weight with respect to the elevator if it started to ascend with the same acceleration?

5

DYNAMIC APPLICATIONS

In this chapter we apply Newton's laws to predict the motions of some simple systems like blocks sliding down inclines and weights attached to pulleys. These are less interesting than the motions of planets and rockets, but even such minor applications show some of the power of understanding nature's laws.

Free Body Diagrams

To analyze the motion (acceleration) of a body, first draw a *free-body* diagram. This means that you isolate the body and sketch all the forces acting *ON* it. You must be careful not to include forces that are exerted by the body on other objects.

We often need to include forces of contact with surfaces like tabletops, walls, and inclines. In the most ideal case where friction can be ignored, the only way a surface can exert force on an object is perpendicular to the surface—this is called a *normal force*.

5.1 Draw a free-body diagram for a 5 N block resting on a horizontal table. (Identify the *normal force* N)

23

5.2   Exercise: Use Newton's second law with acceleration set
       to zero to show that the magnitudes of the normal force
       and the weight are equal for the block of problem 5.1.

Application to Inclined Plane

A favorite object in introductory physics problems is the inclined
plane. Most often, a block is sliding down the plane and we
need to sketch the appropriate free-body diagram.

An obect on a frictionless plane is subjected to only two forces,
gravity acting straight down (*mg*) and the supporting force of the
plane acting perpendicular to the plane (*N*). We are interested in
motion down the plane, so it is helpful to resolve the weight into

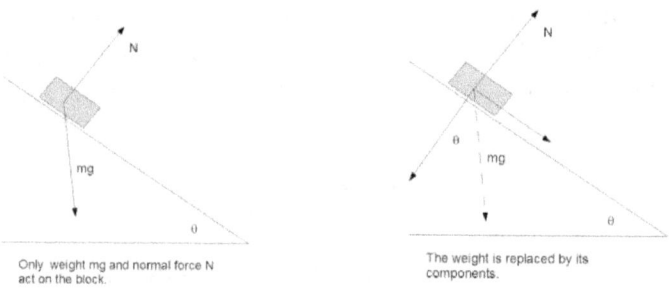

Only weight mg and normal force N
act on the block.

The weight is replaced by its
components.

components—one down the plane and the other perpendicular to
the plane as shown in the diagrams. Note the relation of the
components to the angle θ of the incline.

5.3   Sketch a free-body diagram for a 10 N block sliding
       down a frictionless plane inclined at 30°. Calculate the
       normal force and the component of force acting along the
       plane.

Newton's Second Law applies to both *x*-and *y*-coordinates:

5 DYNAMIC APPLICATIONS

$$F_x = m a_x$$
$$F_y = m a_y$$

We are free to pick these perpendicular directions and it is usually convenient to pick the positive direction to be the direction of motion (or the expected direction of motion) of the body being considered. For instance, for the block of problem 5.2, the positive x-direction would be taken down the plane.

Note: Problems may give mass m or weight w, but not both. One may be calculated from the other by the relation (true on the surface of the earth):

$$w = mg$$

5.4   Calculate the acceleration of 10 N block sliding down a frictionless plane inclined at $30°$.

Note: Many problems are independent of mass. This means that the result is the same, regardless of the value of m and the problem statement will not give a value for m (or w). In these cases, simply use the symbol m (mg for weight) and proceed as though you know the value. The masses will "cancel" at some point in the problem.

5.5   Calculate the acceleration of a block sliding down a frictionless plane inclined at $30°$.

So far we have treated forces due to gravity (weights) and forces due to one surface pressing or supporting another (normal forces). Now we include tension and friction. These embellishments do not change our problem-solving strategy:
(a) draw a detailed free-body diagram,
(b) find acceleration.

Applications with Tension

Definition: *Tension* is the force due to the pull of a rope, cord, or stick. Tension in a rope always *pulls* the object under consideration; an extended rope cannot push an object. Note: The tension T exerted at one end of a rope is also exerted at the other end of the rope, but in the opposite direction.

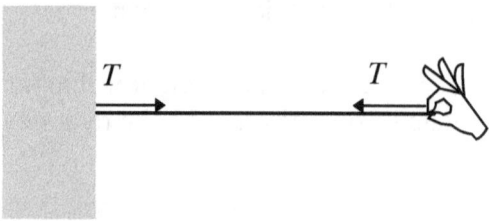

When two objects interact they can be treated separately with their own free body diagrams with $F = ma$ applied to each separately. In general physics problems, this will usually result in two equations in two unknowns. The following problem illustrates this.

5.6   Two 10 kg blocks are joined by a rope and one of the blocks is pulled

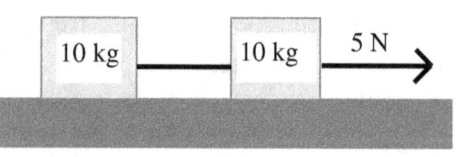

by a second rope with a constant tension of 5 N along a frictionless, horizontal surface. Find (a) the acceleration of the system and (b) the tension in the rope that joins the blocks. Neglect the masses of the ropes.

Here is some help with exercise 5.6. Tension T is shared by both blocks. It points to the right on the left block and to the left on the right hand block. $F = ma$ is then applied to each block. The result is two equations containing unkowns T and a.

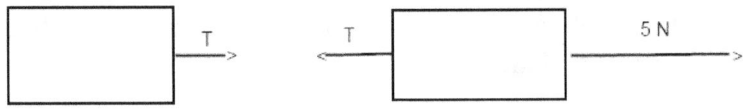

This approach is general and can always be used. However, sometimes simpler approaches will be evident. In the last problem the acceleration could be found immediately by treating both blocks as one 20 kg object pulled by the 5 N force.

5.7   A 10 kg block on a smooth, horizontal table is connected by a cord over a frictionless pulley to a 5 kg mass that hangs over the side of the table. Find the acceleration of the masses and the tension in the cord. (See the figure in problem 5.10. Again, take the direction of acceleration to be positive, even when it is downward.)

Applications with Friction

A force arising from the contact of materials is *friction* and it always opposes the direction of motion. The force of friction, *f*, is usually proportional to the force perpendicular to the surfaces involved; that is, the normal force, *N*. The relevant equation is

$$f = \mu N$$

where the proportionality constant, μ (Greek letter *mu*) is called the *coefficient of friction*. It is a dimensionless quantity.

5.8   A 10 kg block is pushed along a table top by a horizontal force of 5 N. The coefficient of friction between block and table top is 0.03. Calculate the acceleration of the block.

5.9   Calculate the acceleration of a block sliding down a plane inclined at 53°. The coefficient of friction between block and plane is 0.2.

5.10   A 10 kg block on a horizontal table is connected by a cord over a frictionless pulley to a 5 kg mass that hangs over the side of the table. The coefficient of friction between the block and table is 0.2. Find the acceleration of the masses and the tension in the cord.

The friction discussed here is the friction an object experiences when it is moving. This is called *kinetic* friction and the coefficient is often referred to as the *coefficient of kinetic friction.* Friction also occurs between surfaces that do not move relative to each other like the foot of a ladder and the ground it rests on—this is called *static* friction. Static friction is always larger than kinetic friction between the same two surfaces. It takes more force to start moving a heavy trunk along the floor than to keep it moving. Static friction equals whatever force is needed to prevent motion—that is, until a critical force is reached given by $f = \mu_{static} N$ where μ_{static} is the coefficient of static friction.

5 DYNAMIC APPLICATIONS

5.1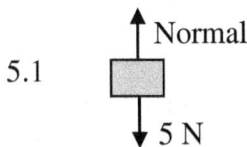

Normal

5 N

5.3   Normal force = 8.7 N, Force down plane = 5 N
5.4 4.9 m/s^2
5.5 4.9 m/s^2
5.6 0.25 m/s^2, 2.5 N
5.7 3.27 m/s^2, 32.7 N
5.8 0.2 m/s^2
5.9 6.6 m/s^2
510   2.0 m/s^2, 39 N

Problems

1.   A heavy particle slides down a smooth inclined plane, starting from rest at the top, the height of the plane being *h* and the length *L*. Find the acceleration of the particle, and the time it will take to reach the bottom.

2.   With what velocity must a particle be projected down a plane 4 m in height and inclined to the horizon at an angle of 30°, so as to reach the bottom in one second?

3.   With what velocity must a particle be projected up a plane 3 m in height and inclined to the horizon at an angle of 30°, so as to reach the top in one second?

4.   The time of falling down a smooth inclined plane is twice that down the vertical height of the plane. Find the ratio of the length of the plane to its height.

5. A 5 kg block on a smooth horizontal table is connected by a cord over a frictionless pulley to a 1 kg mass that hangs over the side of the table. Find the acceleration of the masses and the tension in the cord.

6. A 5 kg block on a horizontal table is connected by a cord over a frictionless pulley to a 1 kg mass that hangs over the side of the table. The coefficient of friction between the block and table is 0.1. Find the acceleration of the masses and the tension in the cord.

7. Masses of 1 and 3 kg hang from the two ends of a cord suspended over a smooth pulley At what speed will they be moving at the end of 1 sec after they are set free?

8. Find the constant tension of a rope which pulls a cart weighing 2000 kg from rest through a distance of 0.6 m up an incline of $30°$ in 2 seconds.

ENERGY

The law of conservation of energy is the most fundamental law of physics. The form of energy conservation presented in this unit can be derived from Newton's laws, but energy conservation is known to be correct even in the domains where Newton's laws are inadequate (the relativistic and quantum-mechanical domains).

Work

When a particle undergoes a displacement s subject to a force F, the *work* W done by this force is defined as the product of the distance moved and the component of F in the direction of motion ($F\cos\theta$, where θ is the angle between F and s).

$$W = Fs\cos\theta$$
$$= \mathbf{F} \cdot \mathbf{s}$$

The notation $\mathbf{F} \cdot \mathbf{s}$ is called the *dot product* of $\mathbf{F}$ and $\mathbf{s}$. Note that the dot product of two vectors is simply the product of the magnitudes and the cosine of the angle between the vectors. The SI unit for work and energy is the *Joule*, J (J=N-m).

When the force varies with each change in displacement Δs, a sum must be used;

$$W = \sum \mathbf{F} \bullet \Delta \mathbf{s}$$

or, in calculus notation,

$$W = \int \mathbf{F} \bullet d\mathbf{s}$$

6.11 A block is pushed 1.5 m along a level surface by a horizontal force of 2 N. The force of friction is 0.6 N. (a) What is the work done by the 2 N force? (b) What is the work done by friction? (c) What is the total work done on the block?

6.12 A 30 N sled is pushed up a frictionless 30° inclined plane by a force of 28 N applied parallel to the ground. The sled is moved 5 m along the plane. Find the work done by (a) the 28 N force, (b) the weight of the sled, (c) the normal force.

Kinetic Energy

The *kinetic energy* of an object is the amount of work the object could perform by virtue of its motion alone. (Or more simply, kinetic energy is the energy due to motion.)

Kinetic energy K for a particle of mass m and speed v is given by

$$K = \tfrac{1}{2}mv^2$$

This represents the maximum amount of work that a particle can perform due to its speed.

6.13 A 3 kg block, initially at rest, is pushed 1.5 m along a level surface by a horizontal force of 2 N. The force of friction is 0.6 N. Compute the (a) acceleration, (b) final velocity, and (c) final kinetic energy of the block. Compare your answer with the total work done in problem 6.1.

Potential Energy

The *potential energy* of a particle is the ideal amount of work that the particle can perform due to its position or configuration. For example, a rock on a roof has the potential ability to drive a stake into the ground below. When the same rock is on the ground, it does not have this ability. Similarly, an expanded rubber band has more ability to do work than when it is not stretched.

There are different equations for potential energy U depending on the nature of the forces including gravity, elasticity, electricity, etc. In this section we consider the potential energy of a mass m at a height y above some "ground" level. The ground level is arbitrary—only differences in potential energies are physically meaningful.

$$U = mgy \quad \text{(applies to masses near Earth's surface)}$$

6.14   A 3 kg block slides 4 m down a plane inclined at 30° with the horizontal. Calculate the block's potential energy difference between the top and bottom of the incline.

Energy Conservation

For friction-free (nondissipative) motion, the law of conservation of energy requires that the sum of kinetic and potential energies never changes. Consequently, when potential energy decreases, kinetic energy must increase to compensate.

6.15   At which labeled point is the cart moving
        (a) fastest, (b) slowest?

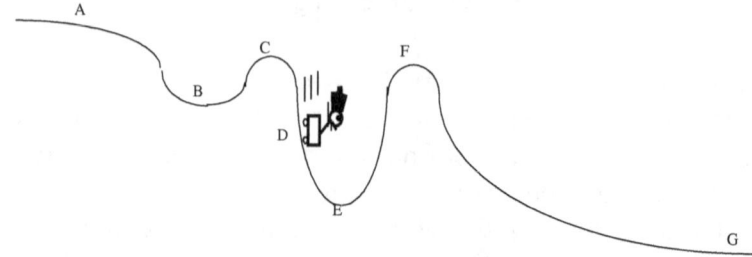

The basic equation for non-dissapative energy conservation is

$$K_1 + U_1 = K_2 + U_2$$

The subscripts refer to any (arbitrary) "initial" and "final" configurations of the system. It is helpful to draw initial and final
situations and record the kinetic and potential energies of each.

6.16   A block slides from rest 4 m down a frictionless plane
        inclined at $30°$ with the horizontal. Use energy conserva-
        tion to find the velocity at the bottom. Note that this
        answer would be the same if the slide were curved.

6.17   A champagne cork shoots straight up out of its bottle at
        8.85 m/s. Find the maximum height it attains above its
        point of release using conservation of energy.

When more than one mass is involved in the motion, the total
kinetic energy K is simply the sum of all the individual kinetic
energies. Likewise, the total potential energy U is the sum of all
the individual potential energies.

6.18   The system in the dia-
        gram is released from
        rest where the 4 kg
        mass is on the floor and
        the 6 kg mass is 0.5 m
        above it. Use conserva-
        tion of energy to find
        the speed with which
        the block strikes the
        floor. Neglect friction and the masses of the cords and
        pulley.

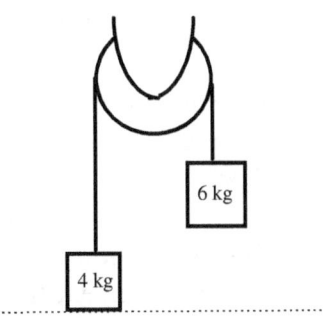

Energy with Dissipative Forces

Friction cannot be included in $K_1 + U_1 = K_2 + U_2$ because there is
no position-dependent potential energy for friction. The following is a more general form of conservation of energy that can
incorporate forces of friction though the work done by friction:

$$W = K_2 - K_1 + U_2 - U_1$$

where W is the work done by any force that is not included in the
U's. W includes friction, but it is not restricted to friction. A
form of this expression that is most often used is the *work-
energy theorem* where *all* forces are included in W and none in
potential energies,

$$W = K_2 - K_1$$

6.19   A 3 kg block, initially at rest, is pushed 1.5 m along a
        level surface by a horizontal force of 2 N. The force of
        friction is 0.6 N. Compute the final velocity of the block
        using conservation of energy.

Elastic Potential Energy

Elastic potential energy may be thought of as the energy stored in a compressed or expanded spring. The elastic force law is *Hooke's law*,

$$F = -k \, x$$

where k is a proportionality constant characterizing the "spring" and x is the displacement from the equilibrium point.. The corresponding potential energy is

$$U_{elastic} = \tfrac{1}{2} k x^2$$

6.20   In an arcade game a 0.1 kg disk is shot across a frictionless horizontal surface by compressing it against a spring and releasing it. If the spring has a spring constant of 200 N-m and is compressed from its equilibrium position by 6 cm, find the speed with which the disk slides across the surface.

Power

Power P is the rate of energy or work expenditure. The defining relation is

$$P = \frac{\Delta W}{\Delta t}$$

The SI units for power are *Watts*, W.

The power expended when a force F acts to move a body at velocity v is

$$P = F v$$

6.21   A student weighing 600 N climbs at constant speed to the top of an 8 m vertical rope in 10 s.Find the average power expended by the student to overcome gravity

6.22   Rob pushes a wheelbarrow by exerting a145 N force horizontally. He moves it 60 m at a constant speed for 25 s.What average power does Rob expend?

Answers

6.1   3 J, -0.9 J, 2.1 J
6.2   121 J, -75 J, 0
6.3   0.47 m/s², 1.18 m/s, 2.1 J
6.4   58.8 J
6.5   G, A
6.6   6.3 m/s
6.7   4 m
6.8   1.4 m/s
6.9   1.2 m/s
6.10   2.7 m/s
6.11   480 W
6.12   33.6 W

Problems

1.   How much work does a 50 kg girl do climbing a mountain 1000 m high?

2.   The lower end of a ladder 16 m long stands on the ground at a distance of 273 cm from a wall against which the upper end rests. How much work will be done in carrying 30 kilos up the ladder?

3.   A pile driver weighs 2000 N. It is lifted 6 m. How much work has been done upon it? How much kinetic energy will it have after falling 6 m. to the pile?

4.   A train weighing 400 tons is moving 30 miles per hour. Compute its kinetic energy. (Change its weight to kilograms and velocity to meters per second.)

5.   What would be the kinetic energy of the train in problem 4 if it were going 60 miles per hour?

6.   A stone of mass 5 kilos is thrown vertically upward with a velocity of 25 m per second. Find its kinetic energy at the end of two seconds.

7.   A champagne cork shoots straight up out of its bottle at 8.85 m/s. Use energy conservation to find the height it attains above its point of release, the velocity with which it returns to its starting point, and its speed when it is halfway to its maximum height.

8.   A rollercoaster is shown in the figure. The coaster car starts from rest at a height $h = 90$ m above the ground. It enters a circular loop with radius $R = 40$ m. Find 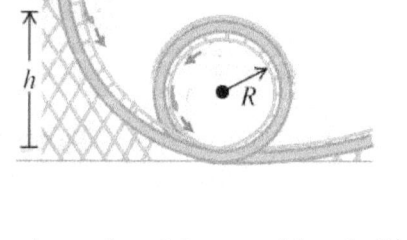 the speed of the car when it reaches (a) ground level, (b) the top of the circle, (c) a level with the center of the circle. Neglect friction.

9.   The coaster car of the last problem is now launched at a speed of 1.0 m/s. Find the speed of the car when it reaches (a) ground level, and (b) the top of the circle.

10. An Atwood machine is depicted in the diagram. Masses m_1 and m_2 are released from rest at the same elevation and m_2 falls through a distance h while m_1 rises by the same distance. Use conservation of energy to derive the following expression for speed v:

$$v^2 = \frac{2(m_2 - m_1)gh}{m_1 + m_2}$$

11. The upper end of a smooth straight wire of length 30 m, is attached to the top of a pole 16 m high. A bead is allowed to slip along the wire from top to bottom. Use energy considerations to find its velocity on reaching the bottom.

PHYSICS LECTURES

MOMENTUM

Momentum in One Dimension

In most of this chapter we are concerned primarily with motion along a line: The *momentum p* of an object of mass *m* and velocity *v* is defined in one dimension as

$$p = mv$$

where the sign of the velocity indicates the direction of motion. The momentum of a collection of objects is the sum of the (signed) individual momenta,

$$p_{total} = p_1 + p_2 + \cdots$$

There is no established unit for momentum, but you can see from the definition (mass times velocity) that is has units of kg m/s.

7.1   Find the momentum of a 50 kg skater with velocity 3 m/s.

7.2   Find the total momentum of two ice skaters; a 75 kg skater with velocity 2 m/s, and a 50 kg skater moving in the same direction at 3 m/s.

Conservation of Momentum

Conservation of momentum requires that the sum of initial momenta equals the sum of final momenta
:

$$\sum p^{initial} = \sum p^{final}$$

This is extremely useful for problems in which there is a collision between objects. Apply conservation of momentum immediately before and immediately after the collision. In solving momentum problems, it helps to sketch initial and final situations.

7.3   A 50 kg girl skates at 1.0 m/s toward a 70 kg boy at rest on his skates. Upon reaching him, she clasps his arm. With what speed do they drift away together?

7.4   A 70 kg boy skates at 1.0 m/s toward a 50 kg girl who is skating away from him in the same direction at 0.5 m/s. Upon reaching her, he holds onto her. With what speed do they drift away together?

Unlike kinetic energy, which is always a positive scalar, the sign of momentum is important. As usual, you can pick any direction to be "positive," but you must be consistent and assign a negative sign to momenta in the opposite direction.

7.5   A 70 kg boy skates at 0.5 m/s toward a 50 kg girl who is skating directly toward him at 1.0 m/s. After colliding, with what velocity (specify direction) does the bruised couple drift away together?

7 MOMENTUM

Events involving the recoil of a gun or an atomic nucleus are the reverse of collisions— particles are flying apart instead of toward each other. These too are treated with conservation of momentum.

7.6   A mobile 500 kg cannon fires a 5 kg cannonball horizontally. The cannon recoils at 2 m/s; find the initial velocity of the cannonball.

Impulse and Momentum

Momentum is *not* conserved when an "outside force" like gravity acts on the system. Imagine that an invisible container encloses the system (colliding skaters, for instance); if a force from outside this container reaches in to affect the system, momentum is not conserved and Eq. (**Error! Reference source not found.**) does not apply. In the case of the colliding skaters, gravity is not a factor because only horizontal motion occurs. Even when two colliding objects exert great forces on each other, these forces are within the imaginary container (system) and momentum conservation applies. However, when an object rises or falls vertically, gravity is an outside force (reaching into the imaginary container) and conservation of momentum does not apply.

The *impulse* delivered to an object subjected to of an average force F for a duration Δt is the product $F \Delta t$. The following expression applies when an outside force causes a momentum change (and momentum is *not* conserved):

$$F \Delta t = p^{final} - p^{initial}$$

7.7   A falling 0.1 kg ball has a velocity of 6.26 m/s just before it hits the floor and then an upward velocity of 5.42 m/s immediately after. The ball is in contact with the floor for 0.01 sec. Find the average force exerted by the floor on the ball. (Notice that the floor exerts the outside force.)

7.8   A ball of mass 0.1 kg falls from a height of 2.0 m and rebounds to a height of 1.5 m. The ball is in contact with the floor for 0.01 sec. Find the average force exerted by the floor on the ball.

Momentum and Energy Combined

In many problems either momentum or energy is conserved only for part of the process. To treat these, we separate the problem into conceptual stages. In the next problem, momentum is conserved immediately before and immediately after the collision, but thereafter gravity is an outside force subject to conservation of energy and not conservation of momentum. It helps to draw the individual stages.

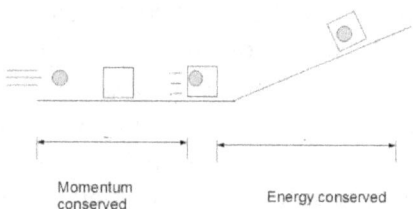

Momentum conserved   Energy conserved

7.9   A 10 g bullet is fired into a 2 kg block at rest that then slides up a frictionless hill to a maximum height of 15 cm above its original position. Find the initial speed of the bullet.

7.10   A 10 g bullet is fired into a 2 kg block at rest that then slides along a rough surface with coefficient of friction 0.5 for 0.6 m before stopping. Calculate the initial speed of the bullet.

7.11   A 10 g bullet is fired at 300 m/s into a 2 kg block at rest that then slides along a rough surface for 0.33 seconds before stopping. Calculate the average force of friction.

7 MOMENTUM

Elastic Collisions

The only collisions where we simultaneously apply both conservation of momentum and conservation of energy are when we are told that the collision is "elastic." A collision between two objects of equal mass (billiard balls) then results in the velocities of the colliding particles being exchanged.

Momentum in Two Dimensions

Momentum is a vector and has vector components:
$$\mathbf{p} = m\mathbf{v} \quad \text{or} \quad p_x = mv_x, \quad p_y = mv_y, \quad p_z = mv_z$$
Note that the fundamental equation for conservation of momentum is a vector relation.

7.12   A 800 kg car skidding due north on a level frictionless icy road at 100 km/h collides with a 1400 kg car skidding due east at 60 km/h in such a way that the two car wrecks stick together and skid as shown in the picture. At what angle θ do the two coupled cars skid?

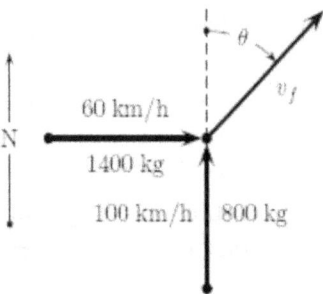

Answers

7.1	150 kg-m/s
7.2	300 kg-m/s
7.3	0.42 m/s
7.4	0.79 m/s
7.5	0.125 m/s in girl's initial direction
7.6	200 m/s
7.7	117 N
7.8	117 N
7.9	343 m/s
7.10	344 m/s
7.11	9.1 N
7.12	46°

Problems

1. A mass of 20 kg moving with a velocity of 18 m/s over-takes a second mass of 32 kg. moving with a velocity of 12 m/s ; find the common velocity after impact. Calculate the loss of kinetic energy for the inelastic collision.

2. A mass of 900 kg moving with a velocity of 30 m/s joins a similar mass moving 10 m/s in the opposite direction. Find the velocity of the combined mass after impact

3. A gun weighing 3,000 kg is placed upon a smooth plane and discharges a 30 kg ball at an elevation of 30°. Find the velocity of the gun's recoil.

4. Carla weighs 55 kg. She sleds from rest at the top of a hill to the ground where she seizes her 15 kg backpack. Her speed is then found to be 10 m/s. How high is the hill?

CIRCULAR MOTION & GRAVITY

The diagram shows a point rotating on a wheel. Its motion is more conveniently described by a radial position r and an angle θ than by (x, y) as we did for projectiles. This is usually true of objects that rotate or orbit. Here we want to develop these angular descriptions and apply them to orbiting objects.

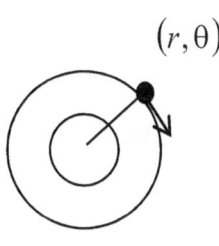

(r, θ)

Circular Motion

In this chapter it is useful to have angles measured in *radians* (the arc length measured in radii). Recall that there are 2π radians in a circle so 1 revolution is 2π radians:

$$1 \text{ rev} = 2\pi \text{ rad}$$

Angular velocity ω (Greek letter omega) is defined as the change in angle per time $\omega = \Delta\theta / \Delta t$ or, in calculus notation,

$$\omega = \frac{d\theta}{dt} \quad \text{in rad/sec}$$

where θ is an angle in radians. *Angular acceleration* α is similarly defined, $\alpha = \Delta\omega/\Delta t$ or, in calculus terms,

$$\alpha = \frac{d\omega}{dt} \quad \text{in rad/sec}^2$$

8.23 A wheel makes 4.5 turns in 2 s. What is its average angular velocity?

Equations for constant angular acceleration are completely analogous to those for linear acceleration. For example, the equation $v = v_0 + at$ has an angular analogue $\omega = \omega_0 + \alpha t$. A useful rule is to begin with the familiar constant acceleration equations and make the following changes:

$$x \rightarrow \theta$$
$$v \rightarrow \omega$$
$$a \rightarrow \alpha$$

8.24 Write the angular analogue equations for the following

$$x = x_o + v_o t + \tfrac{1}{2}at^2$$
$$v = v_o + at$$
$$v^2 = v_o^2 + 2a(x - x_o)$$

8.25 A turntable rotating at 3.5 rad/s slows to a stop in 5 seconds. Calculate the constant angular deceleration.

8.26 What is the angle through which a flywheel turns when it is accelerated at 0.2 rev/s^2 from rest to an angular velocity of 1.0 rev/s?

Connection with Linear Quantities

The angular quantities α, ω, θ are related to their linear counterparts s (arc length), v (tangential velocity), and a (tangential acceleration) by the relations:

$$s = r\theta$$
$$v = r\omega$$
$$a = r\alpha$$

8.27   A truck with wheels of radius 0.5 m travels at 3 m/s. Find the angular velocity of the wheels.

8.28   A truck with wheels of radius 0.5 m travels 10 m. Find the angle (in revolutions) through which the wheels turned.

Period of Rotation

Many problems involving motion in a circle require a link between radius r, tangential velocity v, and the *period* or time for one rotation, T.

$$v = \frac{2\pi r}{T} = \frac{\text{distance around perimeter}}{\text{time to go around circle}}$$

8.29   The earth rotates on its axis once in 24 hours. Find the tangential velocity of an object on the equator. The radius of the earth is 6.38×10^6 meters

Centripetal Acceleration

When a point mass moves in a circle, it must accelerate toward the center–without this *centripetal acceleration* the mass would fly away in a straight line tangent to the circle. Centripetal acceleration exists even when the speed v of the object is constant (because acceleration is a change in either speed or *direction* or both). The basic equation is

$$a_{centripetal} = \frac{v^2}{r}$$

where a is the centripetal acceleration and r is the radius of the circle. The *centripetal force* associated with this acceleration is therefore

$$F_{centripetal} = \frac{mv^2}{r}$$

8.30   The Earth has a mass of 6.0×10^{24} kg and an average orbital radius of 1.5×10^8 km. Calculate the centripetal acceleration a and centripetal force F.

Note that centripetal force is not a "fundamental" force like gravity or electric attraction—it is only the name given to the radial component of such forces. Centripetal force is not drawn as a separate force on a free-body diagram.

8.31   A 0.6 kg rocket propelled toy car has a speed of 4 m/s as it rides inverted at the top of a vertical circular track with radius 1 m. (a) Find the force exerted by the track on the car. (b) What must be the minimum velocity of the car for it to reach the top of the circle?

Gravitation

We have been writing $w=mg$ for the force of gravity on a mass m. However, this is only correct near the surface of the Earth. The force of gravity is a force between any two masses m_1 and m_2 depends on the distance r between the masses:

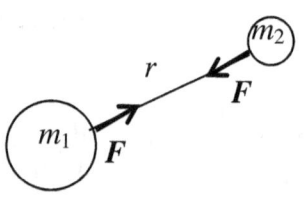

$$F = G\frac{m_1 m_2}{r^2}$$

where the proportionality constant G is a universal constant; that is, it has the same value for any two masses,

$$G = 6.67 \times 10^{-11} \text{ n m}^2 / \text{kg}^2$$

8.32   The mass of the sun is 2.0×10^{30} kg and the mass of the earth is 6.0×10^{24} kg. The distance between the bodies is 1.5×10^8 km. Calculate the force of gravitational attraction and compare it with the centripetal force of problem 8.30.

8.33   Derive the value of g in terms of the mass of the earth M and the radius of the earth R. Evaluate this expression. $M = 6.0 \times 10^{24}$ kg $R = 6.38 \times 10^6$ m.

Gravity is the force that "holds" a planet in orbit around the sun. The gravitational force therefore supplies the centripetal force. Circular motion and gravity, the two topics in this chapter, are brought together in the next problem where gravitational force is set equal to centripetal force. Note that the velocity v can be written in terms of the orbital radius r and the period T using $v = 2\pi r / T$.

8.34   According to Kepler's third law of planetary motion, the ratio T^2 / R^3 is constant, where T is the period and R is the average orbital radius for any planet orbiting the sun. Derive this relation for circular orbits.

Gravitational Potential Energy

We have been writing $U = mgy$ for the potential energy due to gravity, but we must often use a general form that applies away from the surface of the earth. The general gravitation potential energy corresponding to the gravitational force law is

$$U_{gravity} = -\frac{Gm_1 m_2}{r}$$

where r is a distance from the center of the relevant planet or star. Notice that, unlike the force law, this is not an inverse square expression and a negative sign is included.

8.35 Asteroid Toro has a mass of 2×10^{15} kg and a radius of 5×10^3 m. (a) Find the minimum initial velocity needed to throw a stone from the surface of the asteroid in order for it to finally travel infinitely far (escape velocity). (b) Calculate the maximum height to which the stone would rise if it were given the same vertical velocity on earth.

8.36 Challenging question: The approximate expression
$$(1+\varepsilon)^n \approx 1+n\varepsilon$$
is accurate for a number ε much smaller than 1. Let R represent the radius of the earth and let y represent a relatively small elevation from the surface of the earth so that $r = R + y$. Substitute this into
$$U = -Gm_1m_2 / r$$
and use the approximation above to show that
$$U = mgy + \text{constant}.$$
The constant can be ignored because only differences in energies have physical effects.

Answers

8.1   2.25 rev/s or 14.1 rad/s
8.2 $\theta = \theta_0 + \omega_0 t + \frac{1}{2}\alpha t^2$, etc.
8.3   -0.7 rad/s^2
8.4   2.5 rev or 15.7 rad
8.5   6 rad/s
8.6   3.18 rev
8.7   464 m/s
8.8   5.95 X 10^{-3} m/s^2, 3.6 X 10^{22} N
8.9   3.72 N, 3.13 m/s
8.10 3.6 X 10^{22} N
8.13 7.3 m/s, 2.7 m

Problems

1. A body is constrained to move in a circle of 597 cms. radius at a speed of 235 cms. per second. What is the centripetal acceleration and the period of revolution?

2. The distance of the moon from the earth is 3.85×10^8 m and the lunar month is about 27 days 8 hr. What is the centripetal acceleration of the moon toward the earth?

3. Use the data and result of the last problem to calculate the mass of the earth.

4. Two equal masses, A and B, are connected by a string. The mass A describes a circle with radius 1 m and uniform speed on the surface of a smooth horizontal table, while the other mass B is suspended against gravity by the string which passes through a small, smooth hole at the center of the table. Find the speed of A.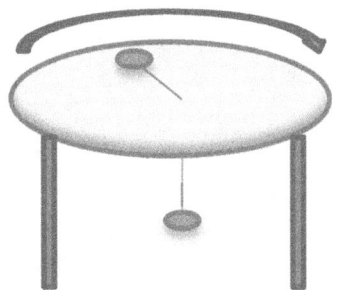

5. A vertical axle revolves 100 times per minute. A rope 1.5 m long is attached to the axle and a mass of 75 kg attached to the other end of the rope is whirled around.. What angle does the cord make with the vertical, and what is the tension in the rope?

6. The length of a conical pendulum is 1 meter, and the radius of the circle in which the bob moves is 60 cm. What is the period of the pendulum?

ROTATIONAL MOTION

The previous unit showed the angular analogues of displacement s, velocity v, and acceleration a. Here the angular analogues of force F and mass m are presented.

Torque

Torque τ is the analogue of force. A torque is defined relative to a specific axis of rotation. (It does not matter if the body actually can rotate about this axis – the axis may even be outside of the body itself.) For the

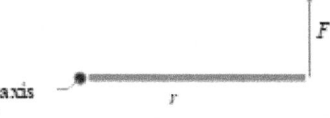

force indicated in the diagram, the magnitude of the force is given by

$$\tau = rF$$

r is called the *moment arm* of the force F. Torque also has a sign indicating the potential direction of rotation; by convention, this is positive for torques that tend to rotate counterclockwise around the axis:

$\tau = \pm r F$

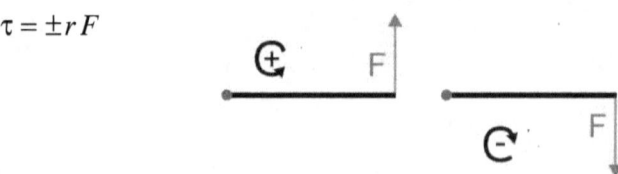

Finally, if the force and the moment arm are not perpendicular, one must take the perpendicular component of either L or F,

$$\tau = \pm r F_{\perp} = \pm r_{\perp} F = \pm r F \sin\theta$$

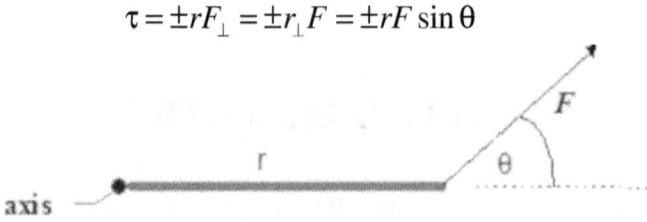

9.1   Find the torque around point P.

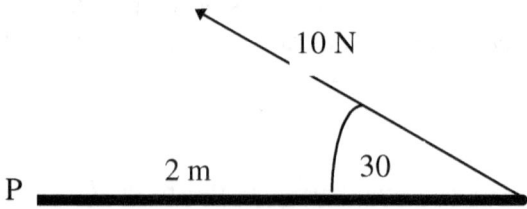

9.2   Find the torque around point P.

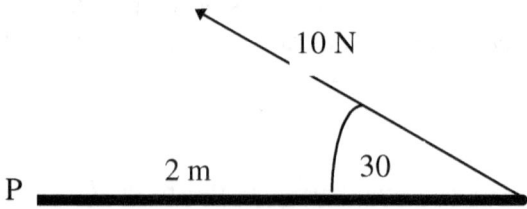

Moment of Inertia

Moment of Inertia I is the angular analogue of mass. The equation for a point-mass m rotating in a circular arc of radius r is

$$I = mr^2 \quad \text{(point mass)}$$

The following are expressions for moments of inertia for two solid objects with mass M:

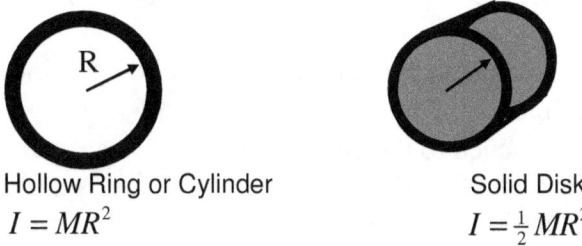

Hollow Ring or Cylinder	Solid Disk
$I = MR^2$	$I = \frac{1}{2}MR^2$

Now the roster of angular analogues is as follows:

$$x \rightarrow \theta \qquad\qquad F \rightarrow \tau$$
$$v \rightarrow \omega \qquad\qquad m \rightarrow I$$
$$a \rightarrow \alpha$$

9.3 Write the angular analogues of the following expressions: $F=ma$ and $K=mv^2/2$.

Rotational Dynamics

Expressions $F=ma$ and its analogue $\tau=I\alpha$ are both used when analyzing forces and accelerations associated with rotational motion.

9.4   A light rope is wrapped around a 3 kg solid cylindrical
       pulley of radius 0.2 m. A force of
       4.9 N is applied to the end of the
       rope.   Calculate the tangential ac-
       celeration of the pulley.

49N

The following problem requires $\tau = I\alpha$ to be
applied to the pulley and $F = ma$ to be applied to the hanging
mass. This results in two equations with two unknowns (acceleration and tension).   Remember to follow the direction of
acceleration as positive, regardless of other conventions.

9.5   A light rope is wrapped around a 3 kg solid cylindrical
       pulley of radius 0.2 m. A mass of
       0.5 kg is attached to the end of the
       rope as shown. Calculate the tangen-
       tial acceleration of the pulley, the
       linear acceleration of the mass, and
       the tension in the rope.

0.5 Kg

The following problem illustrates that the total kinetic energy of
a wheel is the sum of its translational kinetic energy $mv^2/2$ and
its rotational kinetic energy $I\omega^2/2$.

9.6   A ring of radius R is released from rest to roll down a
       rough incline. Calculate its speed when it reaches the
       foot of the incline, $4R$ below the starting point.

Angular Momentum

When no external torque [analogue of force] acts on an object, its angular momentum [analogue of momentum] is conserved:

$$I_1\omega_1 = I_2\omega_2$$

The stick figures show how an ice-skater can execute a rapid spin by first turning slowly with arms and a leg extended to produce a large moment of inertia. When his arms and leg are brought closer to his body, moment of inertia gets smaller and rotation gets faster.

$$I\omega = I\omega$$

9.7   A point mass of 0.05 kg at the end of a cord is rotated in a horizontal circle of radius 0.2 m with an angular velocity of 3 rad/s. The cord is shortened to 0.1 m. (a) Calculate the new angular velocity. (b) Find the change in kinetic energy of the mass. (c) How much work was required to shorten the cord?

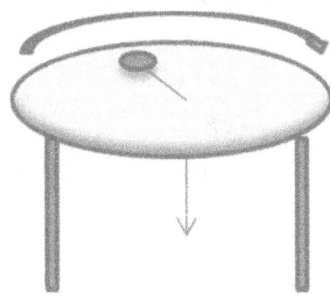

9.8   Advanced Instruction: Angular momentum is best described by a mathematical operation called the *cross
product* (or vector product). Let the vectors **r** and **p** specify the moment arm and instantaneous linear momentum
of a particle respectively. Then angular momentum is a
vector written as $\mathbf{L} = \mathbf{r} \times \mathbf{p}$
where the magnitude of
this operation is

$|\mathbf{L}| = r\,p\sin\theta$ and the direc-

tion of **L** is along the axis
of rotation as shown in the
diagram. Show that for circular motion the magnitude
is given by
(a) $L = mrv$ or (b) $L = I\omega$.

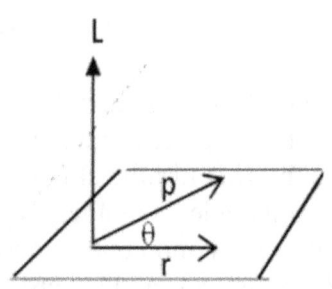

Answers

9.1   −10 N-m
9.2   10 N-m
9.3 $\tau = I\alpha,\quad \frac{1}{2}I\omega^2$
9.4   16.3 rad/s^2
9.5   12.25 rad/s^2, 2.45 m/s^2
9.6 $\sqrt{4gR}$
9.7   48 rad/s, 0,567 J, 0.567 J

Problems

1. A ring-shaped flywheel has radius R and has mass M. If a force F acts steadily upon the wheel at its rim, show that the angular velocity ω after t seconds from the commencement of the motion will be Ft/MR.

2. A ring-shaped flywheel has radius R and has mass M. A mass m is suspended from a light rope wrapped around the wheel's rim. Show that the tension in the rope is given by $mMg/(M+m)$.

3. A homogeneous cylinder of mass M and radius a turns around a horizontal axis coinciding with the axis of the cylinder; a fine thread is wrapped round it, and a mass m is attached to the end. Use energy conservation to show that when the mass m has descended through the height h the angular velocity of the cylinder, neglecting friction, is given by the equation
$$\omega^2 = \frac{4mgh}{a^2(M+2m)}$$

4. A cylinder of radius R is released from rest to roll down a rough incline. Calculate its speed when it reaches the foot of the incline, $5R$ below the starting point. The moment of inertia of a solid cylinder is $\frac{1}{2}MR^2$.

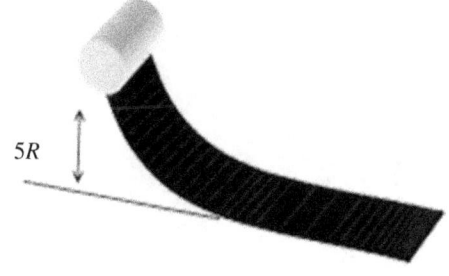

$5R$

5.   A puck on a frictionless air hockey table has a mass of 0.01 kg and is attached to a cord passing through a hole in the surface as in the figure. The puck is revolving at a distance 2 m from the hole with an angular velocity of 3 rad/s. The cord is then pulled from below, shortening the radius to 1.0 m. Find (a) the new angular velocity (in rad/s), (b) the tension in the cord, and (c) the energy required to shorten the radius.

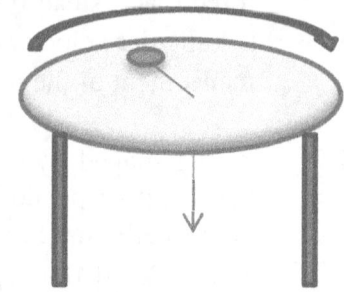

EQUILIBRIUM

The term *equilibrium* refers to objects that do not accelerate. This includes objects that are at rest or moving at a constant velocity.

Equilibrium Conditions

I order for an object to be at equilibrium both linear and angular accelerations must vanish. It follows that the conditions for equilibrium are given by

$$\Sigma F_x = 0, \quad \Sigma F_y = 0, \quad \Sigma F_z = 0$$
$$\text{and}$$
$$\Sigma \tau = 0$$

10.37 Show that the equations for equilibrium follow from expressions of the Second Law.

Center of Mass

The usual task in equilibrium or static problems is to find the unknown forces acting on various parts of an object. It is very convenient to identify a single point at which all the mass of the object may be taken to be concentrated—the *center of mass* (X,Y):

$$X = \frac{\sum_i m_i x_i}{\sum_i m_i}$$

$$Y = \frac{\sum_i m_i y_i}{\sum_i m_i}$$

10.38   A 3 kg plank is 0.6 m in length and has its center of mass at the geometrical center. Masses of 1 kg and 3 kg are attached at opposite ends of the plank. Find the center of mass of the whole system. It should be noted that the center-of-mass is a hypothetical point that may not even be a point on an object. For example, the center-of-mass of a doughnut is somewhere in the doughnut hole.

10.3   Challenging problem: A uniform circular disk of radius R has a circular hole of radius $R/2$ cut out of it as shown. Find the center of mass. (The hole can be treated as a negative mass.)

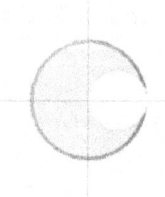

Static Problems

Here is a useful approach to static problems: (i) apply $\sum \mathbf{F} = 0$ to the center-of-mass and (ii) repeatedly apply $\sum \tau = 0$ to various points in the system until you have generated N equations for N unknown forces.

10.4   A horizontal 8 kg plank is 3.0 m in length and has its center of mass at the geometrical center. A 2 kg mass is placed 1.0 m from one end and the plank is supported from the ends. Calculate the forces of support.

10.5   A 20 kg uniform ladder 2.5 meters long rests against a smooth wall at an angle of 53 ° with the horizontal. Calculate the forces on the ladder exerted by the floor and wall.

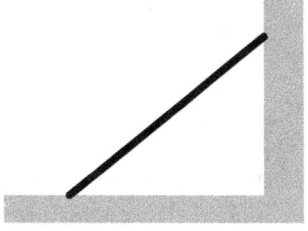

10.6   The diagram shows a uniform 30 cm strut weighing 5 N supporting a 10 N weight 10 cm from the pivot. Find the tension in the cable attached to the wall and find the forces exerted on the strut by the pivot.

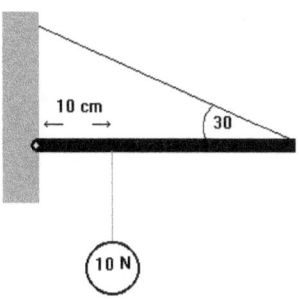

10 cm

30

10 N

Answers

10.2 On the plank, 0.386 m from the 1 kg mass
10.3 A distance $R/6$ to the left of the center along a line joining the radii.
10.4 52.3 N (near the 1 kg mass) and 45.7 N
10.5 196 N (floor), 123 N (wall)
10.6 Tension 11.7 N, Horizontal component of strut 10.1 N to right. Vertical component of strut 9.15 N upward

Problems

1. Two women carry a weight of 80 N slung on a light pole 2.8 m long. If the weight is placed at a distance of 1 m from one end, what weight does each woman carry?

2. A stick of timber of uniform cross section is carried by three men, one at one end and two by means of a bar placed crosswise under the stick. Where must the bar be placed that each man may carry one-third the weight?

3. A horizontal bar AB, 3 meters long, has one end B attached to the vertical side of a building. The other end A is supported by a rope tied to it and to the building at a point 4 meters above B. The bar supports a mass of 50 kg at its middle point. Find the tension in the rope (by moments).

FLUID MECHANICS

The term *fluid* applies to a liquid or gas. Although Newton's laws of motion apply to fluids, new concepts like density and pressure often replace mass and force as useful measures.

Density

The *density* of an object ρ (Greek letter *rho*) is given by the expression

$$\rho = \frac{m}{V}$$

where m is the mass of the object and V is its volume. Standard International (SI) units of density are therefore kg/m^3, but it is much more commonplace for it to be given in g/cm^3 (or gm/cc). Equation (1) defines *mass density*; occasionally a *weight density* is given (these can be identified by units of force/volume such as lb/ft^3). The density of water is a special standard:

$$\rho_W = 1 \text{ gm/cm}^3$$
$$= 1000 \text{ kg/m}^3$$

11.39 Find the side of a 172 g brass cube (density 8.6 g/cm^3).

Pressure

Pressure is defined as force per unit area,

$$P = \frac{F}{A}$$

There are more standard units for pressure than any other physical quantity. The more important of these are

1 Pa = 1 N/m^2 ,   1 Torr = 1 mm Hg, 1 atm = 760 Torr
1 bar = 10^5 Pa ,   1 atm = 1.013×10^5 Pa = 1.013 bar

11.40 A cylindrical storage tank has a base area of 0.784 m^2 and contains 4004 kg of water. Find the pressure at the bottom of the tank in Pascals (Pa) and atmospheres (atm). Ignore pressure due to the atmosphere.

Pressure Below a Fluid

Pressure increases with depth below the surface of a fluid. The pressure due to the weight of fluid above is called *gauge pressure* P_G, and is calculated by

$$P_G = \rho g h$$

where h is the depth below the surface and ρ is the density of the fluid. We often need to find the *total pressure P* at this point by adding the ambient or atmospheric pressure $P_{ATM.}$ to the gauge pressure,

$$P = P_{ATM} + P_G$$

Atmospheric pressure is due to the weight of air above. There is a standard atmospheric pressure of 1.013 Pa in MKS units, but the actual atmospheric pressure may be slightly higher or lower on any given day.

11.41 At what depth below water is the gauge pressure equal to the standard atmospheric pressure? What is the total pressure at this depth (in atm)?

A contiguous fluid at rest must have equal pressures at equal levels. This fact is useful in solving many problems. The following problem can be solved without equations simply by reading pressures as millimeters of mercury and then applying this principle.

11.42 Mercury rests in an open-tube manometer as shown. Atmospheric pressure (actual pressure, not standard 760 Torr) is 750 Torr. Find (a) the gauge pressure at the bottom of the U-tube, (b) the absolute pressure at the bottom of the tube's left arm, (c) the pressure of the gas in the tank.

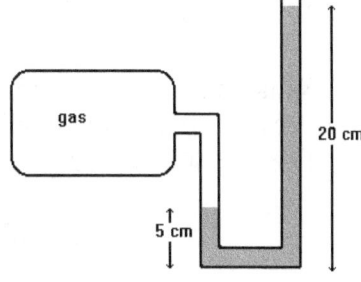

Archimedes' Principle

Archimedes' principle addresses buoyancy in fluids: A body immersed in a fluid is buoyed up by a force equal to the weight of the fluid that is displaced by that body. In equation form, the *buoyant force* is given by

$$F_B = \rho_{Fluid} V_{Submerged} \, g$$

where ρ_{Fluid} is the density of the fluid (*not* the object in the fluid) and $V_{Submerged}$ is the volume of the submerged portion of the object (not the volume of the object unless it is entirely below the fluid surface).

Buoyancy problems are treated by drawing a free-body force diagram for the object in the fluid. When the object is in equilibrium, the sum of the vertical forces, including the buoyant force (acting upward), must equal zero.

11.43   Under water, it takes 2 N of force to lift a 12 N turtle (it is said to have an *apparent weight* of 2 N) when under water. Calculate the volume of the turtle in cm^3.

11.44   A block of wood with density 0.6 g/cc.floats in water. What percent of the volume of the block is below water?

11.45   What must be the volume of a slab of floating ice in order to hold a 1000 kg polar bear without wetting her feet? Density of ice=0.92 g/cc. Density of sea water, 1.03 g/cc.

Pascal's Rule

A pressure increase in part of a contiguous fluid is communicated uniformly thoughout the fluid. In the following problem, the pressures on both ends of the hydrolic lift are equal.

11.8   The small piston of a hydraulic lift has a cross-sectional area of 10 cm^2 and the large piston has an area of 200 cm^2, as in the figure. What force F must be applied to the small piston to raise 80 kN?

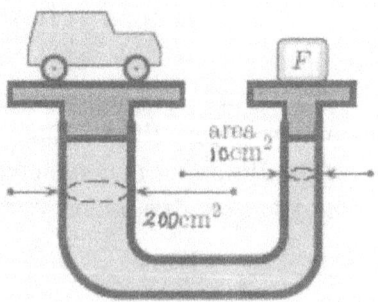

Fluid Flow Rate

When fluid flows through a pipe of cross sectional area *A* with velocity *v*, the volume flow rate *dV/dt* is given by

$$\frac{dV}{dt} = Av$$

For steady flow in a pipe or hose, the volume flow rate must be the same throughout, even with varying cross section and flow velocity:

$$A_1v_1 = A_2v_2$$

11.9 Calculate the average velocity of the blood in an aorta of radius 1.0 cm if the flow rate is 5 liter/min (1 liter = 1000 cc).

11.10 A hose of area 4 cm^2 carries fluid at 30 cm/s. An obstruction in the hose narrows the area to 3 cm^2. What is the fluid velocity in the obstructed portion?

Bernoulli's Equation

Bernoulli's equation expresses conservation of energy for a moving fluid. It relates fluid pressure and flow velocity at two points in a moving fluid.

$$P_1 + \tfrac{1}{2}\rho_1 v_1^2 + \rho_1 g h_1 = P_2 + \tfrac{1}{2}\rho_2 v_2^2 + \rho_2 g h_2$$

In the following standard problem we use this equation to relate an arbitrary point on the surface of a tank to an exit point below. We are given a "large" tank so that the velocity at the surface can be neglected. Pressures at the surface and at the exit point are taken to be equal atmospheric pressures.

11.11 A large open tank is filled with water. A small hole is made in the side of the tank 50 cm below the top. Find the velocity with which the stream of water leaves the tank.

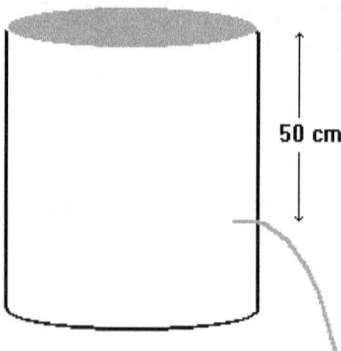

50 cm

Answers

11.1	2.7 cm
11.2	5×10^4 Pa, 0.5 atm
11.3	10.3 m, 2 atm
11.4	200 Torr or mm Hg, 950 Torr, 900 Torr
11.5	1020 cm^3
11.6	60%
11.7	9.1 m^3
11.8	4 kN
11.9	26.5 cm/s
11.10	40 cm/s
11.11	3.1 m/s

11 FLUID MECHANICS

Problems

1. An airplane in level flight whose mass is 20,000 kg has a wing area of 60 m^2. What is the pressure difference between the upper and lower surfaces of its wing? Express your answer in atmospheres.

2. Find the minimum gauge pressure needed in a water pipe entering at ground level to provide water to a floor 30 m above.

3. What is the pressure in atmospheres at a depth of 100 meters in water?

4. A basketball floats in a bathtub of water. The ball has a mass of 0.5 kg and a diameter of 22 cm.
   (a) What is the buoyant force?
   (b) What is the volume of water displaced by the ball?

5. What must be the volume of a helium balloon in order for it to lift a load of 200 kg? The density of air is 129 kg/m^3 and the density of helium is 0.18 kg/m^3. Neglect the weight of the balloon fabric, but include the weight of helium.

6. Water at a gauge pressure of 3.8 atmospheres at street-level flows into an office building in the speed of 0.60 m/s through a pipe 5 cm in diameter. The pipe tapers down to 2.6 cm in diameter by the top floor 20 m above. Calculate the flow velocity and the gauge pressure in the pipe on the top floor.

7.   The velocity of flow across the top of an airplane wing is 320 m/s and the velocity across the bottom is 300 m/s. The wing has an area of 20 m². Make the minor approximation that the top and bottom of the wing are approximately at the same level. The density of air is 1.3 kg/m³.
(a) Find the difference in pressure between the top and the bottom of the wing in Pa.
(b) What is the upward force exerted on the wing?

$$12$$

THERMAL PHYSICS

The concepts of temperature and heat are related, but they are not the same things. A glass of ice water at a tropical picnic may have a temperature of 0°C. It will stay at this temperature as long as some ice remains. When the ice has all melted the water temperature will rise until it matches the warm temperature of the surroundings. One way to appreciate this is to view heat as a form of energy and temperature as a measure of molecular motion. Heat energy flows into the ice water but this energy goes into breaking the bonds of the ice. Only when all the ice is turned to liquid water does the heat energy go into molecular motions and thereby increase the temperature.

Temperature

The *absolute* or *Kelvin* scale of temperature has a lowest possible value of 0 (absolute zero). This scale of temperature is the most fundamental as it is related directly to molecular motions. The Kelvin temperature T_K is related to the centigrade or Celsius temperature T_C by

$$T_K = T_C + 273.15$$

The units of Kelvin temperature are kelvins or K.

Materials tend to expand when heated. Two expressions of this fact are

$$\Delta L = \alpha L \Delta T$$

where α (*alpha*) is called the *coefficient of linear expansion*, and

$$\Delta V = \beta V \Delta T$$

where β (*beta*) is called the *coefficient of volume expansion*. The relation between α and β is

$$\beta = 3\alpha$$

12.1   A metal rail is 800.00 m long at 30°C and 800.12 m long at 40°C. Find the coefficients of linear expansion and volume expansion in this temperature range.

Ideal Gas

Fluids have a temperature T, pressure P, volume V, and these quantities are interrelated. A sample of ideal gas relates P, V, and T through the ideal gas equation:

$$PV = nRT$$

where n is the number of moles given by

$$n = \frac{m}{M} = \frac{\text{mass of sample}}{\text{molecular mass}}$$

and R is a universal constant of proportionality called the *gas constant*:

$R = 8.314$ J/mol-K
$R = 1.99$ cal/mol-K
$R = 0.082$ L-atm/mol-K

12.2   A 25-L tank contains 64 gm of oxygen at 27°C. What is the number of moles of oxygen and what is the pressure in the tank?

When a sample of gas changes form state 1 to state 2, it is convenient to use the following form:

$$\frac{P_1 V_1}{n_1 T_1} = \frac{P_2 V_2}{n_2 T_2}$$

12.3   A bubble, originally 2 cm^3 in volume, rises from the bottom of a lake to the surface where the pressure is 1 atm and expands to 6 cm^3. The temperature at the bottom is 7°C and the temperature at the surface is 27°C. Calculate the pressure at the bottom of the lake.

Heat

Heat is a form of energy and is measured in Joules or calories. In particular, heat is the form of energy transferred to a system solely because of a temperature difference.

$$1 \text{ cal} = 4.186 \text{ J}$$

A material of mass m undergoes a temperature change ΔT when it labsorbs an increment of heat ΔQ:

$$\Delta Q = cm \, \Delta T$$

The proportionality constant c is called *specific heat* and is different for different substances. The specific heat of water is

$$c_w = 1 \text{ cal/g-°C} = 4190 \text{ J/kg-K}$$

12.4   How much heat (in calories) is needed to raise the temperature of 200 gm of water from 20°C to 30°C

12.5 A lead bullet moving at 300 m/s embeds in a tree. As-
suming all the energy is converted into heat in the bullet,
what is its temperature rise?
Specific heat lead=130 J/kg-K.

Phase Changes

Freezing and boiling are examples of phase changes where a material changes form to a solid, liquid, or vapor (gas). Phase
changes for a particular material occur at a specific temperature
and pressure; water freezes at 0°C at 1 standard atmosphere of
pressure. This temperature does not change throughout the
transformation and $\Delta Q = cm\,\Delta T$ does not apply. During a phase
change,

$$\Delta Q = m\,L$$

Here L is the *heat of fusion, heat of vaporization*, or *heat of sub-
limation*. Heat of fusion applies to both freezing and melting
but the appropriate sign must be used; +heat for melting, -heat
for freezing. A similar remark applies to vaporization and con-
densation.

12.6 How much heat (in calories) is required to melt 1 g of ice
at 0°C ? (heat of fusion of water=80 cal/g).

12.7 How much heat (in calories) is required to heat 1 g of ice
at -10°C to steam at 100°C? (specific heat ice=0.5 cal/g-
K, heat fusion of water=80 cal/gm, heat of vaporization
of water=540 cal/gm)

12 THERMAL PHYSICS

Calorimetry

When objects at different temperatures are put together in an insulated container, the sum of heat transfers must be zero.

$$\sum_{\text{all substances}} \Delta Q = 0$$

where ΔQ is given by $\Delta Q = cm\,\Delta T$ or $\Delta Q = m\,L$.

12.8 An aluminum can of mass 0.5 kg contains 0.118 kg of water at a temperature of 20°C. A 0.2-kg block of iron is dropped into the can. The final temperature of the mix is 25°C. Find the initial temperature of the iron.
sp heat Al=910 J/kg-K, sp. heat Fe=470 J/kg-K,
sp. heat water=4190 J/kg-K

Answers

12.1	1.5×10^{-5}, 4.5×10^{-5}
12.2	2, 1.97 atm
12.3	2.8 atm
12.4	2000 cal
12.5	346°C
12.6	80 cal
12.7	725 cal
12.8	75°C

Problems

1.   A glass flask holds 200 c.c. of water at 0^0C. How much will it hold at 100°C? The coefficient of linear expansion for glass is 0.0000083.

2.   The density of a piece of silver at 0^0C is 10.5. Find its density at 100°C if its coefficient of cubical expansion is 0.0000583.

3.   The volume of a mass of copper at 50°C is 500 cc; find its volume at 300°C. Coefficient of cubical expansion, 0.0000565.

4.   How much ice at 0^0C will be melted by 30 grams of copper (specific heat, 0.095) at 200°C?

5.   A mass of 100 gms. of platinum (specific heat, 0.0355) is heated in a furnace and is then dropped into 200 grams of water at 0^0C; the temperature of the water rises to 26° C. What was the temperature of the furnace?

6.   If a kilo. of copper at 100° C. be placed in a cavity in a block of ice at 0°C., and if 119 gms. of ice are melted, find the heat of fusion of ice.

7   100 gms. of ice at - 20' C. were thrown into 1 kilo. of water at 20° C. contained in a copper vessel weighing 100 gms. When the ice was nielted the temperature of the water was 10°.15 C. Find the heat of fusion of ice.

8.   A vessel is filled with a gas at 15° C. and a pressure of 100 mm of mercury; find the pressure at 100° C.

13

SIMPLE HARMONIC MOTION

When dynamics problems involve constant accelerations, we use expressions like $v = v_0 + at$. In general, however, forces are variable and the familiar constant acceleration equations are not applicable. Oscillatory motion presented here is a very important category of variable acceleration.

Hooke's Law

Hooke's law gives the elastic force F on an object that is displaced a distance x from its equilibrium position:

$$F = -kx$$

where k is the *spring constant*.

13.46 A spring scale with spring constant 20 N/m is elongated 20 cm. What is the weight on the scale?

Period and Frequency

An object experiencing the force in Eq.(**Error! Reference source not found.**) oscillates in a regular fashion called *simple harmonic motion* (SHM). Such an object returns to its original position in a definite time T called the *period* of the motion. The object's *frequency* f is the number of times per second that it returns to a specific position. f is measured in $\sec^{-1}$ or *Hertz*, Hz.

The relation between period T and frequency f is the same as for circular motion,

$$T = \frac{1}{f}$$

The defining relation for *angular frequency* ω is

$$\omega = 2\pi f$$

Springs and simple pendula represent oscillatory systems and the following expressions for their angular frequencies are usually memorized:

$$\text{Spring} \quad \omega = \sqrt{\frac{k}{m}}$$

$$\text{Pendulum} \quad \omega = \sqrt{\frac{g}{L}}$$

13.47 A 0.5 kg bob is attached to a spring with spring constant 200 N/m. Find the angular frequency, the frequency, and the period of the motion.

13.48 Find the length of a simple pendulum that keeps time with a 2 second period.

Simple Harmonic Motion Equations

The following are the basic equations of SHM shown with their constant acceleration analogs. The *amplitude A* is the maximum displacement from equilibrium and the *phase angle* θ_0 is usually determined from initial conditions.

Simple Harmonic Motion	Constant a Analog
$a = -\omega^2 x$ ↑ The defining equation for SHM	a = constant
$x = A\cos(\omega t + \theta_0)$	$x = v_0 t + \frac{1}{2}at^2$
$v = -\omega A \sin(\omega t + \theta_0)$	$v = v_0 + at$
$v^2 + (\omega x)^2 = (\omega A)^2$	$v^2 = v_0^2 + 2ax$

13.49   A 0.5 kg object is suspended from a spring with spring constant, k = 5 N/m, and moves with simple harmonic motion. At the instant that it is displaced from equilibrium by –20 cm, what is its acceleration?

13.5   An object oscillates with angular frequency 10 rad/s after it is released from rest at a displacement of 50 cm. What is its speed when the displacement is 40 cm?

13.6   An object that vibrates with simple harmonic motion is released from rest at a displacement of 15 cm and oscillates with a frequency of 4 Hz. Compute the
(a) maximum magnitudes of velocity and acceleration,
(b) velocity and acceleration when x is 9 cm,
(c) time to move from equilibrium to a x=12 cm.

The mathematics of harmonic motion is used to model the vibrations of molecules, oscillations of light waves, sound waves, and many other fundamental processes.

13.7   Calculus based: (a) Show that $x = A\cos(\omega t + \theta_0)$ is a solution to the defining equation $a = -\omega^2 x$. (b) Use the result from (a) to show $v = -\omega A \sin(\omega t + \theta_0)$. (c) Use conservation of energy to show $v^2 + (\omega x)^2 = (\omega A)^2$.

Answers

13.1   4 N
13.2   20 rad/s, 3.18 Hz, 0.314 s
13.3   1 m
13.4   2 m/s^2
13.5   3 m/s
13.6   (a) 3.77 m/s, 94.7 m/s^2 (b) 3.0 m/s, 56.8 m//s^2 (c) 1.54 s

Problems

1.   A spring oscillates with angular frequency 10 rad/s after it is released from rest at a displacement of 25 cm. What is its speed (in cm/s) when the displacement is 20 cm?

2.   A swing is suspended from a tree limb at the end of a light rope 4 m long. The swing is released from rest at a displacement of 25 cm from the vertical..
Find the frequency of the swing in Hz.
What is the maximum velocity of the swing in m/s?
How many seconds (minimum) does the swing take to move from one end of its oscillation to the other?

3.   A 0.2 kg object is suspended from a spring with spring constant, k = 10 N/m, and moves with simple harmonic motion. At the instant that it is displaced –0.05 m from equilibrium what is its acceleration in m/s2?

4.   An object oscillates with a spring constant of k = 15 N/m. What is the potential energy of the system when the object displacement is 0.04 m?

5.   Find the period of oscillation of a pendulum 6 meters long at a place where g is 9,8 m/s2.

6.   If the length of the seconds pendulum is 99.41 cm, what is the value of g?

7.   A seconds pendulum is lengthened 1 per cent. How much does it lose a day?

\

WAVES

Waves are readily seen in disturbed water and vibrating strings, but the wave concept is fundamental to interpreting sound, light, atoms and molecules, and properties of all fundamental particles. This chapter gives a first exposure to the concept and waves are treated again in more detail in chapter 18.

Wavelength, Frequency, & Amplitude

A sinusoidal wave is characterized by a wavelength λ as shown in the diagram. For a traveling wave, the period T of the wave is the time it takes for a point on the wave to move a distance of one wavelength. The speed c of the wave is therefore given by

$$c = \frac{\lambda}{T}$$

It is more usual to speak in terms of wave *frequency* rather than the period. The frequency f is the number

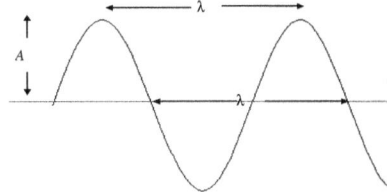

87

of wavelengths that pass per second. The units were formerly called cycles-per-second, but the modern unit is simply *Hertz*, Hz

$$f = \frac{1}{T}$$

The expression $c = \lambda/T$ is then written as

$$c = \lambda f$$

14.50 All light waves travel (in vacuum) with a speed of 3×10^8 m/s. Find the frequency and period of blue light with a wavelength of 4.5×10^{-7} m.

The amplitude A is a measure of the size or intensity of the wave. In the language of simple harmonic motion, amplitude is the "A" in the expression $x = A \cos \omega t$.

14.51 Indicate the amplitude and wavelength of the wave shown in the figure.

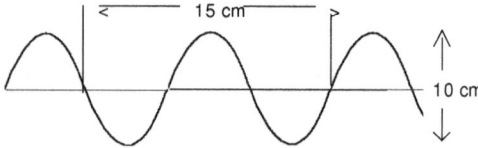

Speed of Waves in Strings

The speed v of a wave on a string depends on the tension F and the *linear density* or mass per unit length μ of the string:

$$v = \sqrt{\frac{F}{\mu}}$$

14.52 Find the speed of a wave on a wire of linear density 10^{-3} kg/m under a tension of 160 N. What is the frequency of a 2.0 m wave on this wire?

Doppler Effect

An object that emits a wave is called the *source* of the wave. For example, a loudspeaker is a source of sound waves and a broadcast antenna is a source of radio waves.

When the source of a wave is moving, the waves in front of the source are crowded together and the waves in back of the source are spread out as shown below. Since the wavelengths in front of the source are shorter than when the source is stationary, the frequency is higher. Similarly, the frequency is lower than the stationary frequency behind the moving source.

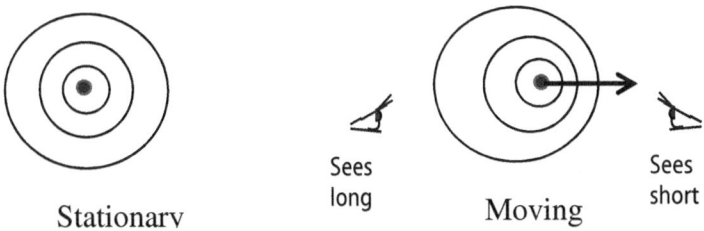

Stationary

Sees long

Moving

Sees short

14.53 A distant train whistle is higher pitched than that of a stationary train. Is the train approaching or receding? Explain.

14.54 A star is speeding away from the earth. Is the light from the star more red or more blue than the light from a similar star at rest relative to us? Explain.

These are qualitative considerations. A mathematical description of the Doppler effect is given in chapter 18.

14 WAVES

Superposition

Waves can "add" together to make a wave of greater amplitude or they can "cancel" each other (or some intermediate result between these two extremes.) These effects are called constructive and destructive interference.

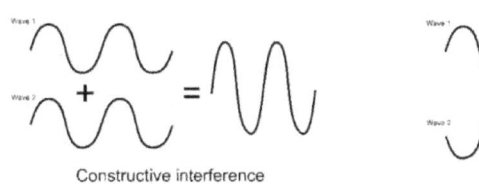

Constructive interference   Destructive interference

Answers
14.1 6.7×10^{14} Hz, 1.5×10^{-15} s
14.2   5 cm, 10 cm
14.3   400 m/s, 200 Hz
14.4   approaching, as shorter waves have higher frequency
14.5   more red, as red wavelengths are longest in the visible spectrum

Problems

1.   What is the wavelength of a 60 megahertz (60×10^{6} Hz) wave broadcast by a FM radio station?

2.   The velocity of light in water is about $0.75c$ where c is the speed of light in vacuum. Wave frequency, however, is the same for water or vacuum so that respective wavelengths must differ. Find the wavelength in water for the 60 megahertz wave of problem 1.

15

ELECTROMAGNETISM AND LIGHT

The science of electromagnetism produced a great synthesis —the four Maxwell equations that govern the subject were found to describe not only all electrical and magnetic phenomena, but all of radiation including radio waves, radar, x-rays, light and optics, etc. A full mathematical treatment of this synthesis is beyond the this course, but this chapter describes it qualitatively.

Charges and Coulomb's Law

Two kinds of charge are known to exist, positive charges and negative charges. All charges come in (positive or negative) multiples of e, the charge of an electron or proton.

Description of Coulomb's Law:
Charges of the same sign repel each other and charges of opposite sign attract each other. Coulomb's law states this fact mathematically and, given the amounts of the charges and the distance between them, it predicts the forces exerted on the charges.

15.1   *True or False?*
(a) Electrons attract each other.
(b) The attraction between an electron and a proton increases as the distance between them increases.

The Field Concept

The question arises- How does a charge "know" of another charge's existence so that it can exert the proper force in the proper direction when these charges do not touch, see, smell, or hear each other! This is called the action-at-a-distance problem. It can be resolved by asserting that the charges *do* touch each other through an invisible disturbance in space that connects the charges. This illusive connection is called a *field.*

The *electric field* is a disturbance in the space surrounding a charge that can exert a force on other charges. (the field can be thought of as a kind of extension of the charge). A charge is therefore called a *source* of the electric field.

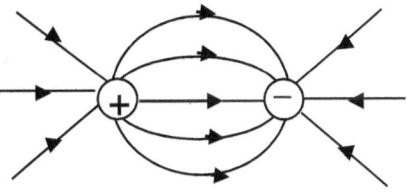

The Electric Field Causes Charges to Exert Forces on Each Other.

We will see that an electric field can exist even without a source under certain conditions. The electric field is denoted as **E.**

Note: Many other fields are known to exist including magnetic fields and gravitational fields.

Note: One of the four Maxwell equations is Coulomb's Law written in terms of the electric field.

Notice that the density of field lines decreases with increasing distance from the charge. (That is, the strength of the field decreases with distance.)

15.2 *True or False?*
   (a) An electric field acts as an intermediate agent by which charges interact.
   (b) An electric field acts as a disturbance in the space surrounding a charge.
   (c) Electric fields exert forces on charges.

Currents and Magnetic Fields

Definition:
When charges move (as they do in a wire), the flow is called a *current*.

Currents exert forces on other currents. These forces are attributed to a *magnetic field* denoted as **B.** Magnetic fields do not exert forces on stationary charges; they do exert forces on moving charges.

A magnetic field surrounds a current and this field exerts a force on other currents.

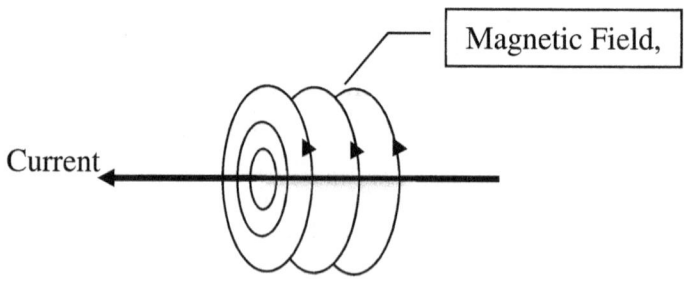

A Current Causes a B field
to Circulate Around it

15.3   *True or False?*
   (a) Moving charges comprise a current.
   (b) A magnetic field surrounds a current.
   (c) Magnetic fields exert forces on currents.
   (d) Magnetic fields exert forces on stationary charges.

Ampere-Maxwell Law

The fact that a current causes a B-field to arise is part of a more complete statement called the Ampere-Maxwell law. This law recognizes that a B-field will also arise to circulate around a *changing* E-field The electric field may change by becoming stronger or weaker or by changing direction.

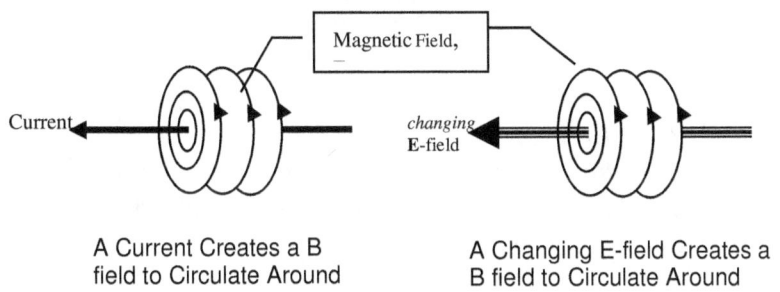

A Current Creates a B   A Changing E-field Creates a
field to Circulate Around   B field to Circulate Around

Ampere-Maxwell Law: **B**-fields arise to circulate around currents or around changing E-fields.

A static electric field does not create a B-field.

15.4   *True or False?*
   (a). A *changing* electric field is surrounded by a B-field.
   (b) A *static* electric field is surrounded by a B-field

Faraday's Law

It is possible to generate an E field without charge. A magnetic field that is changing by becoming stronger, weaker, or by changing direction causes an E-field to circulate around it.

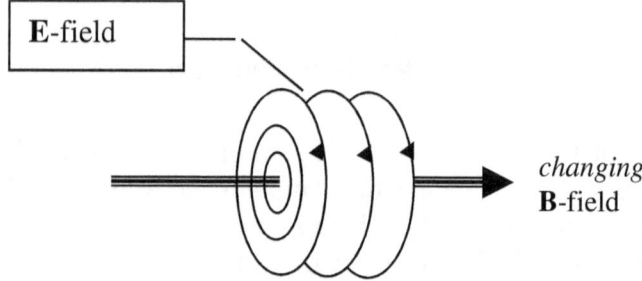

A Changing B-field Causes an E field
to Circulate Around it

15.5 *True or False?*
   (a) A *changing* magnetic field generates an E-field.
   (b) A *static* magnetic field is surrounded by an E-field.

No Magnetic Monopole

Magnetic fields are generated by currents or changing E-fields, but there is no stationary "magnetic charge" or **magnetic monopole** that has an attached B-field in the way an electric charge has an attached E-field. This seemingly negative statement has physical content and is one of the four Maxwell equations.

A qualitative summary of the four Maxwell equations follows in the next section.

15 ELECTROMAGNETISM AND LIGHT

Maxwell Equations

Coulomb Law:
> Charges are a source for E-fields

Magnetic Monopole
> There is no elementary source for B-fields

Faraday's Law
> Changing B-fields cause surrounding E-fields

Ampere-Maxwell Law
> Changing E-fields cause surrounding B-fields and
> currents cause surrounding B-fields

Radiation -- the Fields are Real!

Maxwell asked the question, "Are the fields real?" He did this by looking for solutions to his equations with all charges and currents removed. If E-and B-fields can persist without these, the fields must have an independent existence.

He found that the fields can exist without charge or current in the form of waves that must move with the enormous speed of 3×10^8 m/s , the speed of light. He concluded that light is a wave of electric and magnetic fields!

Although the wave solutions to Maxwell's equations demand that all electromagnetic waves travel at the speed of light c, any wavelength (distance between wave crests) is allowed. Since the human eye can see only a limited range of wavelengths (roughly 380 nm to 760 nm), electromagnetic theory predicts that light has cousins that we cannot see. All these waves are referred to as *electromagnetic radiation*.

You can get an idea of how the waves are propagated by starting with a changing B-field (possibly initiated by sending a current through an antenna wire) and then applying Faraday's law to create a changing E-field which, by the Ampere-Maxwell law, induces a changing B-field, etc.

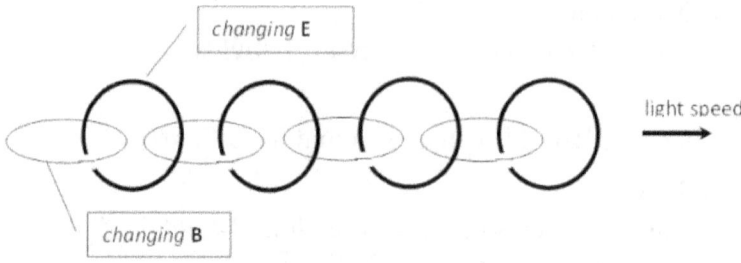

15.6   *True or False?*
   (a). Light consists of E-and B-fields
   (b)   Electromagnetic radiation can propagate through space without charges or currents.
   (c)   In electromagnetic waves, changing E fields and changing B fields alternately produce each other

The Electromagnetic Spectrum

The list in this section gives a bare mention of the more familiar forms of electromagnetic waves.

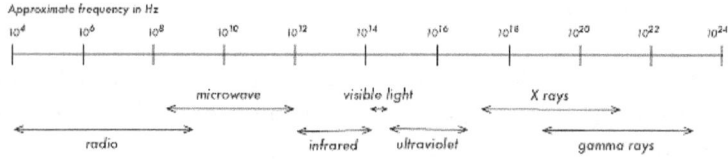

Radio: This long-wave radiation includes radio and television broadcasts and emissions from certain stars.

Microwave and Radar: These frequencies are easily reflected from metals. Microwave frequencies readily stimulate water molecules to move so that food with significant water content is cooked rapidly in microwave ovens.

Infrared: "heat waves" are radiated from hot objects. Actually, infrared is radiated from all objects, but it is more abundant from objects with higher temperature. Almost half of sunlight is in the infrared regions.

Visible Light: Visible wavelengths range from about 380 nm (violet) to 760 nm (red). A mnemonic device to remember the colors of the spectrum in order of decreasing wavelength is the name ROY G. BIV.

Ultraviolet: Kills bacteria and tans skin. Shorter wavelengths can cause skin cancer. About 7% of sunlight is in the ultraviolet and higher frequencies. Much of this is absorbed by ozone (O_3) in the upper atmosphere.

X-rays: Easily penetrate flesh (less so for bones).

Gamma rays: (γ-rays) These occur in nuclear explosions and nuclear decays.

15.7 List the following in order from lowest to highest frequency: orange light, blue light, infrared, x-rays, microwave, radio.

Answers
15.1   F, F
15.2   T, T, T
15.3   T, T, T, F
15.4   T, F
15.5   T, F
15.6   T, T, T
15.7   radio, microwave, infrared, orange light, blue light, x-rays

16

THE NATURE OF LIGHT

Speed of Light

All electromagnetic waves travel (in vacuum) at the speed c,

$$c = 3 \times 10^8 \text{ m/s}$$

and obey the following relation between wavelength λ and frequency f:

$$c = \lambda f$$

16.1   Find the wavelength of a 1200 MHz radio wave.

Reflection & Refraction

The diagram below depicts a ray of light being partly reflected and partly refracted by a transparent medium like glass or water. The angles obey the *law of reflection* and the *law of refraction*. Notice that the angles are measured from the normal to the interface.

$$\theta_1 = \theta_3 \qquad \text{law of reflection}$$

$$n_1 \sin \theta_1 = n_2 \sin \theta_2 \qquad \text{law of refraction}$$

The index of refraction, n, is a characteristic of the medium; the greater the ability to bend light, the larger the value n. Unless otherwise stated, the index of refraction of air is taken to be 1, the same as the index for a vacuum.

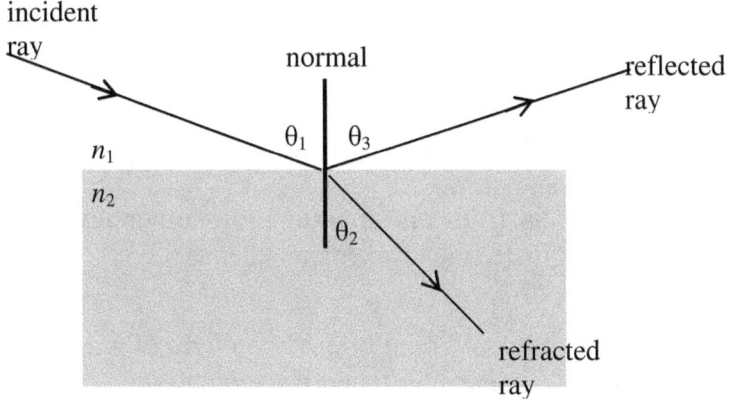

Reflection and Refraction

16.2   Light is Incident on water ($n = 1.33$) at an angle of 60°. Find the angle of refraction in the water and sketch rays for the refraction and the associated reflection.

Although light travels at speed c in a vacuum, it is slower in a material medium The index of refraction n is related to v, the speed of light in the material, by the expression

$$n = c/v$$

16.3   What is the speed of light in glass with index of refraction 1.5?

The term *total internal reflection* is given to a phenomenon where no refraction can occur so the rays are completely reflected. This can occur for rays that move from a medium of larger to smaller refractive index. The *critical angle* θ_c above which only total reflection can occur is given by

$$\sin \theta_c = \frac{n_{small}}{n_{large}}$$

16.4   A point source of light is 1.0 m below the surface of a pool of water ($n=1.33$). Find the radius of a circle of light seen by people outside the pool.

When a wave travels from one medium to another, its wavelength changes, but its frequency remains unchanged.

16.5   Derive the following relation between light of wavelength λ_{medium} in a medium and wavelength λ_{vacuum} in vacuum:

$$\lambda_{medium} = \frac{\lambda_{vacuum}}{n}$$

Answers

16.1   0.25 m
16.2 40.6^0
16.3 2×10^8 m/s
16.4   1.14 m

16 THE NATURE OF LIGHT

Problems

1.   Light is incident on glass with refractive index 1.5 at an angle of 30°. Find the angle of refraction in the glass.

2.   What is the speed of light in a diamond (index of refraction 2.4)

3.   A fiber optic cable with refractive index 1.50 is submerged in water (refractive index of water is 1.33). Find the critical angle above which the light stays in the cable.

4.   Light from a point source under water fills a circle of radius 1.5 m at the surface. The refractive index of water is 1.333. How far below the surface is the light?

17

THIN LENSES

Objects reflect or emit light rays in all directions. A lens can catch some of these rays and bend (refract) them to construct an image of the original object.

Thin Lens Equation

The thin lens equation relates an *object distance s* to an *image distance s'* and a characteristic of the lens called the *focal length f*. Given any two of these quantities, the third can be calculated from the thin lens equation,

$$\frac{1}{s} + \frac{1}{s'} = \frac{1}{f}$$

Two kinds of thin lens are depicted here, *converging* and *diverging*.

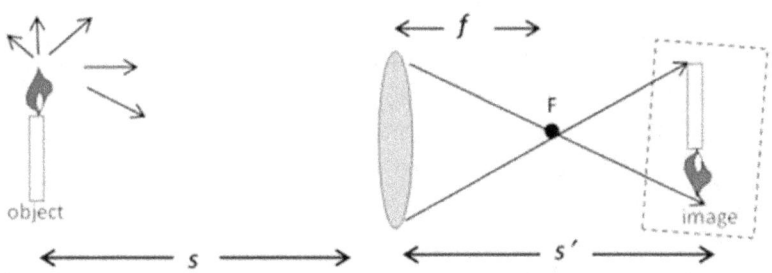

Converging Lens

F is the *focus* where all parallel rays converge. This diagram shows an image that can be captured on a screen, a *real image*.

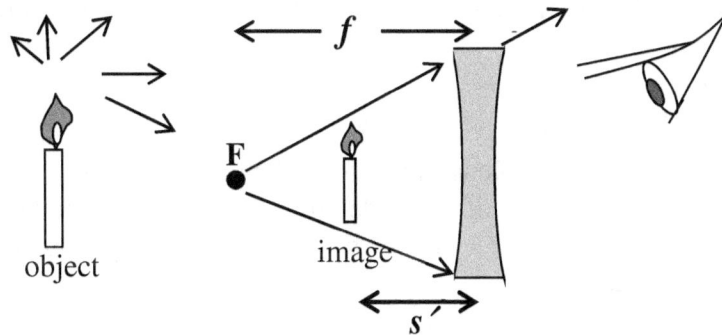

Diverging Lens

F is the focus from which all parallel rays appear to diverge. This diagram shows an image that can be seen as shown but not captured on a screen, a *virtual image*

Sign Conventions

(a)   When s' is positive it signifies a *real image* that can be projected on a screen.
         When s' is negative it signifies a *virtual image* made from diverging rays.

(b)   When *f* is positive it signifies a
      converging lens (convex).
      When *f* is negative it signifies a
      diverging lens (concave).

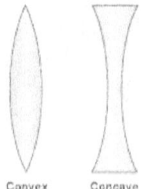

Convex   Concave

(c)   The object distance *s* is positive except when it is on the
      "wrong" side of the lens. That is, the side opposite that
      which gathers light. This only occurs when the "object"
      is an image from another lens.

17.1   A converging lens has a focal length of 20 cm. For an
       object distance of 60 cm, determine (a) the distance of
       the image from the lens and (b) whether the image is real
       or virtual.

17.2   A converging lens has a focal length of 20 cm. For an
       object distance of 6 cm, determine (a) the distance of the
       image from the lens and (b) whether the image is real or
       virtual.

Ray Sketching

Two intersecting rays are used to locate an image point. the simplest rays to draw are (i) a straight ray through the optical center
and (ii) a ray that is directed parallel to the optical axis and is
then deviated through the focus. The following diagrams show
the two most usual ray drawings. Note that rays from the base of
an object pass right through the optical axis; this is the reason we
like to draw an object with its base situated there.

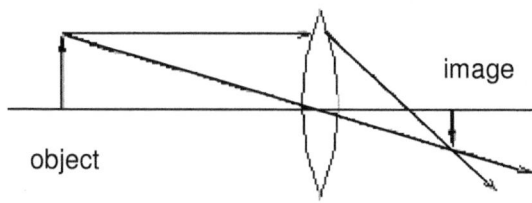

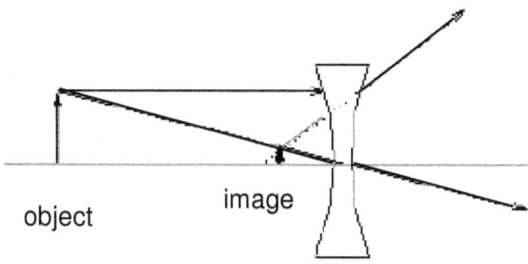

17.3   Sketch ray diagrams for the object in the last two problems. Identify the images and determine, from the
diagrams alone, whether they are (i) real or virtual and
(ii) erect or inverted.

Magnification

When magnification m is larger than 1, the image is larger than
the original object. Often, it is smaller than 1, indicating a reduced image. The equation given below uses a sign convention
that indicates when the image is upright or inverted.

$$m = -\frac{s'}{s}$$

The sign convention: when m is
   positive $\rightarrow$ image upright
   negative $\rightarrow$ image inverted

17.4   Repeat problem 17.1, this time indicating the magnification and whether the image is upright or inverted. Compare the results with the diagrammatic approach of problem 17.3. If the original object was 3 cm tall, how tall is the image?

17.5   Repeat problem 17.2, this time indicating the magnification and whether the image is upright or inverted. Compare the results with the diagrammatic approach of problem 17.3. If the original object was 3 cm tall, how tall is the image?

17.6   A diverging lens has a focal length of 20 cm. For an object distance of 20 cm, determine (a) the position of the image, (b) whether the image is real or virtual (c) the magnification, and (d) whether the image is upright or inverted. Draw a ray diagram.

Spherical Mirrors

Spherical mirrors obey the thin lens equation (1), the magnification equation (2), and the ray diagram rules apply as well. Here the sign of f for a concave (converging) mirror is taken to be positive and f for a convex (diverging) mirror is taken to be negative. All other sign conventions hold for spherical mirrors as they do for lenses.

Here is a relation between the mirror's radius of curvature R and the focal length:

$$f = \pm \frac{R}{2} \quad \left\{ \begin{matrix} converging \\ diverging \end{matrix} \right\}$$

17.7   A hubcap has a focal length of 10 cm. Find its radius of curvature.

Corrective Lenses & Vision

Nearsightedness or *myopia* is the inability to see distant objects clearly. It is characterized by a far point, which is the farthest distance at which the eye sees clearly. To correct the vision to see father, find a lens with a focal distance f that will take an object at infinity ($s = \infty$) and form an image at the far point ($s' = -$far pt). Eyeglasses must form virtual images (no screen in front of the eyes), so s' is negative.

17.8   What is the focal length of a lens that corrects nearsighted vision when the far point is 20 cm? Is the lens converging or diverging?

Farsightedness or *hyperopia* is the inability to see nearby objects clearly. It is characterized by a near point, which is the closest distance at which the eye sees clearly. By convention, the accepted close distance at which the corrected eye should see clearly is 25 cm. To correct the vision, find a lens with a focal distance f that will take an object at 25 cm ($s = 25$ cm) and form an image at the near point ($s' = -$ near pt). Again, eyeglasses must form virtual images (no screen in front of the eyes) so s' is negative.

17.9   What is the focal length of a lens that corrects farsighted vision when the near point is 50 cm? Is the lens converging or diverging?

Lens Prescriptions

Finally, we note that a corrective lens is not usually referred to by the focal length, but by $1/f$, called the *power* of the lens and measured in *diopters* or reciprocal meters:

$$\text{power in diopters} = \frac{1}{f \text{ (in meters)}} = \frac{100}{f \text{ (in cm)}}$$

17.10 What power lens corrects (a) nearsighted vision with a far point of 20 cm and (b) farsighted vision with a near point of 50 cm?

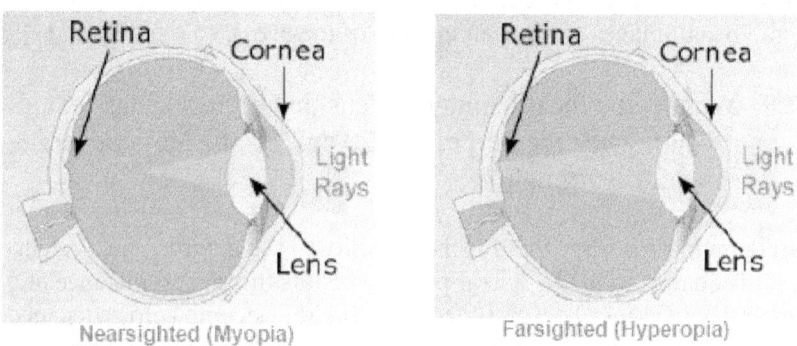

Nearsighted (Myopia)

Farsighted (Hyperopia)

The focus forms before the retina in nearsightedness and behind the retina in farsightedness

Lens Combinations

To find the image from two lenses aligned in sequence, we use the image of the first lens to serve as the object of the second lens. Often the image from the first lens falls behind the second lens. In such cases, treat the object distance as negative for the second lens.

17.11 The following four multiple choice questions relate to the following system. Two thin lenses, each with a local length of 20.0 cm are aligned on a common axis and are separated by 30 cm. A small object is placed on the axis, 40 cm to the left of the first lens.

1. A relation between object distance s, image distance s', and focal length f is
   (a) $\dfrac{1}{s} - \dfrac{1}{s'} = \dfrac{1}{f}$, (b) $\dfrac{1}{s} + \dfrac{1}{s'} = f$, (c) $\dfrac{1}{s} + \dfrac{1}{s'} = \dfrac{1}{f}$,
   (d) $\dfrac{1}{f} + \dfrac{1}{s'} = \dfrac{1}{s}$, (e) $s - s' = f$

2. Where would the Image form in relation to the first lens if this were the only lens? (a) 20 cm to the right, (b) 40 cm to the right, (c) 20 cm to the left (d) 40 cm to the left (e) 6.67cm to the right

3. The object distance s to be used for the second lens is
   (a) –10 cm, (b) +10 cm (c) 20 cm, (d) -6.67 cm, (e) 6.67 cm

4. The final Image distance for both lenses in relation to the second lens is
   (a) -10cm, (b) I 0cm (c) 20cm, (d) -6.67cm, (e) 6.67cm

Answers

17.1   30 cm, real

17.2   8.57 cm, virtual

17.4   - ½ , inverted, 1.5 cm

17.5   1.43, upright, 4.3 cm

17.6   -10 cm, virtual, ½, upright

17.7   -20 cm

17.8   -20 cm, diverging

17.9   50 cm, converging

17.10 -5 diopters, 2 diopters

17.11 c, b, a, e

Problems

1.   A diverging lens has a focal length of 30 cm. For an object distance of 20 cm, find the distance of the image from the lens.

2.   A candle flame is placed at a distance of 30 cms. from a converging lens with focal length 7.5 cms. Find the position of the image. Is it erect or inverted?

3.   A candle flame is placed at a distance of 30 cms. from a convex mirror made from a sphere of 30 cms, diameter, Find the position of the image. Is it real or virtual? Is it erect or inverted?

4.   What power lens (in diopters) corrects nearsighted vision with a far point of 50 cm?

5.   Ms. Green uses +1 diopter reading glasses to read comfortably at 25 cm. What is the nearest distance at which she reads unaided?

WAVES REDUX

This chapter reviews and extends our treatment of waves to prepare for considering the physical properties of light and for introducing quantum physics.

Wavelength, Frequency, & Amplitude

A sinusoidal wave is characterized by a wavelength λ as shown in the diagram. For a traveling wave, the period T of the wave is the time it takes for a point on the wave to move a distance of one wavelength. The speed c of the wave is therefore given by

$$c = \frac{\lambda}{T}$$

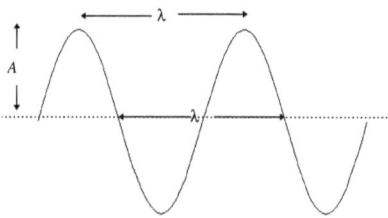

It is more commonplace to speak in terms of wave *frequency* rather than the period. The frequency *f* is the number of wavelengths that pass per second. The units were formerly called cycles-per-second, but the modern unit is simply *Hertz*, Hz

$$f = \frac{1}{T}$$

The expression for velocity is then written as

$$c = \lambda f$$

18.1   All light waves travel (in vacuum) at a speed of 3×10^8 m/s. Find the frequency and period of blue light with a wavelength of 4.5×10^{-7} m.

The amplitude *A* is a measure of the size or intensity of the wave. In the language of simple harmonic motion, amplitude is the "*A*" in the expression $x = A \cos \omega t$. Notice that the amplitude is not the full height of the wave; it is the distance from the midpoint. Also note that each wavelength consists of a peak and troth or two "lumps."

18.55   Indicate the amplitude and wavelength of the wave shown in the figure.

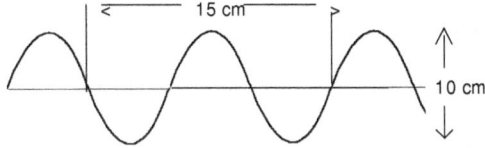

Speed of Waves in Strings

The speed v of a wave on a string depends on the tension F and the *linear density* or mass per unit length μ of the string:

$$v = \sqrt{\frac{F}{\mu}}$$

18.3   Find the speed of a wave on a wire of linear density 10^{-3} kg/m under a tension of 160 N. What is the frequency of a 2.0 m wave on this wire?

Wave Mathematics

The shape of a stationary wave can be written as $y = A\sin\dfrac{2\pi}{\lambda}x$ so that the shape repeats as x increases by λ. When a wave moves to the right with velocity v, it travels a distance vt in time t. The shape remains the same if vt is subtracted from x,

$$y = A\sin\left[\frac{2\pi}{\lambda}(x - vt)\right].$$

Most often, this is written in the form

$$y = A\sin(kx - \omega t)$$

where k is the *wave number* and ω is the angular frequency,

$$k = 2\pi/\lambda \quad \text{and} \quad \omega = 2\pi/T = 2\pi f$$

From $v = \lambda f$, it is easy to show

$$v = \omega/k .$$

18.4   Comparing the last two expressions for y, show that $k = 2\pi/\lambda$ and $\omega = 2\pi f$.

18.5   Given a wave $y = 0.3\sin(3x - 7t + 0.2)$, find (a) the velocity of the wave, (b) the wavelength, (c) the frequency, (d) Calculus based: the maximum amplitude of the *transverse* speed of the wave.

Doppler Equation

Consider a sound source moving with speed v_s relative to an observer. The frequency f' heard by the observer is

$$f' = \frac{f}{1 \pm \dfrac{v_s}{v}}$$

where the "+" sign is for the source moving away from the observer, f is the frequency for a stationary observer and v is the velocity of sound.

18.6   The frequency of a train whistle is known to be 442 Hz. The frequency heard by a person near the tracks as the train approaches is 446 Hz. How fast is the train moving? The speed of sound in air is 345 m/s.

Answers

18.1 6.7×10^{14} Hz, 1.5×10^{-15} s
18.2   5 cm, 10 cm
18.3   400 m/s, 200 Hz
18.5 $v = 7/3$, $\lambda = 2\pi/3$, $f = 7/(2\pi)$, 2.1
18.6   3.09 m/s

19

INTERFERENCE

Two waves having the same wavelength can "add" together to make a wave of greater amplitude or they can "cancel" each other (or some intermediate result between these two extremes.) These effects are called *constructive and destructive* interference.

Constructive and Destructive Interference

Constructive interference occurs when the paths of the two waves differ by an integer number (m) of wavelengths (λ). These waves are said to be "in phase." In this case peaks combine with peaks and troughs combine with troughs. Destructive interference occurs when the paths of the two waves differ by a half-integer number ($m+\frac{1}{2}$) of wavelengths. These waves are "out of phase" and peaks and troughs combine to cancel each other.

$$\text{Path difference} = m\lambda \quad \text{Constructive Inteference}$$

$$\text{Path difference} = (m+\tfrac{1}{2})\,\lambda \quad \text{Distructive Interference}$$

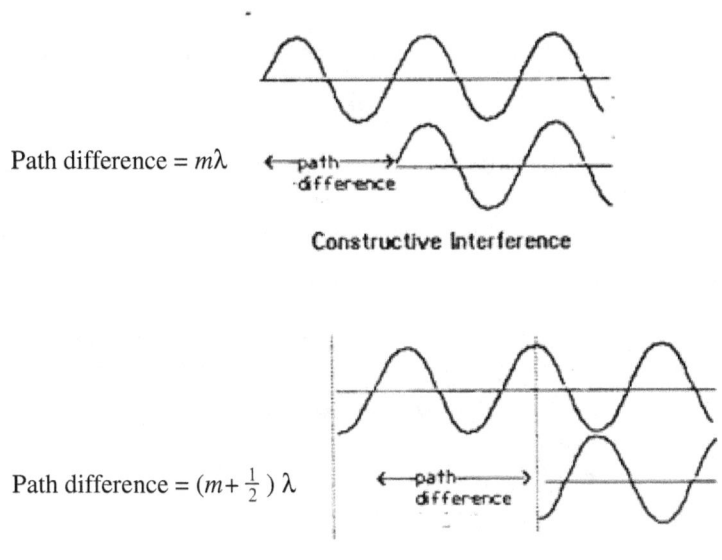

Path difference = $m\lambda$

Constructive Interference

Path difference = $(m+\frac{1}{2})\lambda$

Destructive Interference

Double Slit Interference

Young's double slit experiment is a favorite application of the theory of interference. Light with a single wavelength λ is incident on a card with two parallel slits as shown on the following figure. The light then falls on a screen showing-not two bands-but numerous bands or fringes.

Light source   Slits   Screen

.The basic equation for the bright fringes of light on the screen is

$$d \sin\theta = m\lambda$$

Here d is the distance between slits and θ is the angle from the central axis to the mth bright fringe as shown below. All bright fringes are approximately the same distance apart and the dark fringes are equally spaced between the bright fringes.

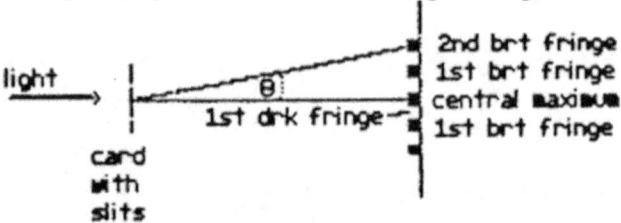

Young's Double Slit Experiment

Note:
Many problems relating to the double slit experiment are made easier when you recognize that, for small angles θ (say $5°$ or less), the following approximations hold:

$$\sin\theta \simeq \tan\theta \simeq \theta \text{ (in radians)}.$$

Note also that wavelengths are often given in *nanometers*, nm:

$$1 \text{ nm} = 1\times10^{-9} \text{ m}$$

19.1   In a double slit experiment, the third bright fringe is 15 mm from the central fringe. What is the distance of the first dark fringe from the central maximum?

19.2   In a double slit experiment, the slits are spaced 0.2 mm apart and the screen is 2 m away. The third bright fringe is 15 mm from the central fringe. Find the wavelength of the light.

Thin Films

Another familiar application of interference regards reflection of light from thin films. Reflection occurs at top and bottom surfaces of the film. When the reflected waves meet, constructive or destructive interference can occur and cause a bright (maximum) reflection or dark reflection (minimum).

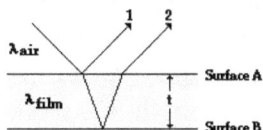

Thin film problems are treated by calculating the path differences between the two reflected waves and then setting this equal to $m\lambda_{\text{film}}$ for maximum reflection or $\left(m+\frac{1}{2}\right)\lambda_{\text{film}}$ for minimal reflection. Most often you are asked to find the minimum film thickness in either case. Then choose the smallest integer m that gives a positive thickness t. Here are three things to note in calculating the path difference Δpath :

- The wave makes a "round trip" in a film of thickness t, so this causes a path difference $2t$ between the exiting waves. $\Delta\text{path} = 2t + \cdots$

- Each reflection from a material of higher index of refraction adds a half wavelength $\lambda_{\text{film}}/2$ to the original path. Reflection from a less dense material does not shift the path length.
  $\Delta\text{path} = 2t + \lambda_{\text{film}}/2$ for each reflection from low to high n

- The wavelength of a ray changes upon entering a material (but frequency does not change). The wavelength inside a film λ_{film}, is related to the vacuum wavelength λ and refractive index n by $\lambda_{\text{film}} = \lambda/n$

19.3   Soap film with refractive index 1.4 is used to make a bubble. What is the thinnest soap film that will maximally reflect red light of wavelength 750 nm?

19.4   Glass of refractive index 1.5 is coated with a film of material having refractive index 1.4.
(a) What is the thinnest coat that will maximally reflect red light of wavelength 750 nm?
(b) What is the thinnest coat that will maximally absorb red light of wavelength 750 nm?

19.5   Glass of refractive index 1.5 is coated with a film of material having refractive index 1.2. What is the thinnest coat that will not reflect (absorb) blue light of wavelength 450 nm?

Single Slit Diffraction

A narrow single slit can cause a fringe pattern to form on a screen similar to the double slit pattern. This is due to different parts of the wave self-interfering. The basic equation for the *dark* fringes is

$$w\sin\theta = m\lambda$$

where w is the width of the opening. This looks like the double slit equation, but it is important to note that it applies to minima or dark fringes.

19.6   A slit is illuminated by light of wavelength 589 nm and a diffraction pattern is formed on a screen 3.0 m away. If the distance from the central maximum to the first dark fringe is 1.77 mm, find the slit width.

19.7   A slit 1.0 mm wide is illuminated by light of wavelength 589 nm. A diffraction pattern forms 3.0 m away. Find the distance from the central maximum to the first dark fringe.

Diffraction Gratings

When waves are incident normally on a diffraction grating, bright bands form behind the grate at an angle θ from the normal given by the expression

$$d \sin \theta = m\lambda \qquad d = \text{distance between slits}$$

Often, the number of slits per centimeter, N, is given rather than d. The reciprocal of N gives the number of cm per slit:

$$d = \frac{1}{N}$$

19.8   A diffraction grating has 10^4 parallel slits per centimeter. Monochromatic light incident normally has a first order bright fringe at $30°$. What is the wavelength of the light?

Answers

19.1   2.5 mm
19.2 5×10^{-7} m or 500 nm
19.3   134 nm
19.4   125 nm, 250 nm
19.5   93.75 nm
19.6   1 mm
19.7   1.77 mm
19.8   500 nm

Problems

1. Light with a wavelength of 646 nm passes through two
    slits and forms an interference pattern on a screen 8.75 m
    away. The distance between the central bright fringe and
    the first-order ($m = 1$) bright fringe is 5.16 cm.
    (a) What is the separation between the slits?
    (b) What will be the distance between the central bright
    fringe and the second-order ($m = 2$) *minimum*?

2. Light reflected from a thin film of oil ($n = 1.40$) floating
    on water ($n = 1.33$) constructively interferes at a wave-
    length of $\lambda = 550$ nm.
    (a) Sketch the situation and indicate which of the reflect-
    ed rays undergo a phase shift.
    (b) Find the minimum thickness of the film that could
    produce this constructive interference.

19 INTERFERENCE

20

COULOMB'S LAW

Coulomb's law describes the forces between charges. This is the primary force responsible for the structure of atoms and molecules (analogous to gravity as the primary force responsible for the solar system and other star systems). The Coulomb force determines most of chemistry.

Charge

Electric charges are recognized by the forces they exert on each other. There are two kinds of charge, positive and negative. Charges of similar sign repel each other and those of opposite sign attract. The unit of charge in the MKS system Is called the *Coulomb* and Is written as "C."

Charge is quantized. That is, it comes in elementary 'lumps" that are indivisible. The basic lump is e, the charge magnitude of an electron (electrons have charge $-e$).

$$e = 1.60 \times 10^{-19} \text{ C}$$

Use a periodic table to identify X in the exercise that follows. Remember that the charge number is in the left subscript and the number of nucleons (protons plus neutrons) is in the left superscript. Charges and nucleons must be conserved. Thus the sum of subscripts on the left of the reaction must match the sum on the right. Similarly, the sum of superscripts must match.

20.1   Identify the element denoted by X
$$^{35}_{16}S \rightarrow ^{0}_{-1}e + X$$

Coulomb's Law

The force exerted by charge q_1 on q_2 and vice-versa has a magnitude F given by

$$F = k\frac{q_1 q_2}{r^2}$$

with the universal coefficient k given by

$$k = \frac{1}{4\pi\varepsilon_0} \simeq 9\times10^9$$

.

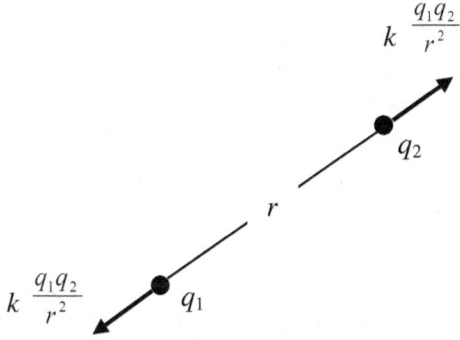

The direction of this force is along the line segment (r) joining the point charges. The force points away from charges of similar sign and toward charges of opposite sign.

20.2 A point charge of 6µC (µC = 1 *micro* Coulomb = 10^{-6}C) is separated by 5 cm from a charge of –2µC. Find the force on each charge. Are the forces attractive or repulsive?

When a charge q is influenced by several other charges, the force on q is the vector sum of the individual forces due to each of the other charges.

20.3 Three charges are arranged along a straight line as shown. Calculate the magnitude and direction of the Coulomb force on each charge.

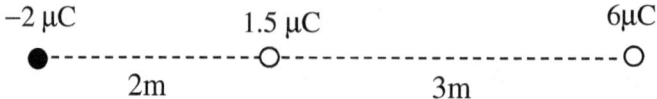

−2 µC   1.5 µC   6µC

2m   3m

20.4 Four charges are arranged at the corners of a square 3 m on a side as shown in the diagram. Calculate the magnitude and direction of the Coulomb force on a charge of 1.0×10^{-4} C located at point P.

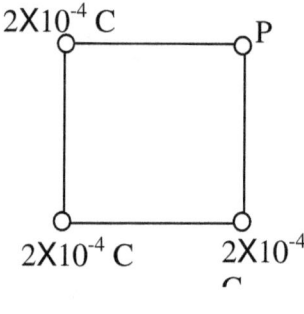

2×10^{-4} C   P

2×10^{-4} C 2×10^{-4} C

20.5 Two identical charges of 1.0×10^{-7} C are separated by 1.0 cm.Calculate the force magnitude between the charges .

20.6 Two identical charges of 1.0×10^{-7} C are separated by 1.0 cm.What is the force when the separation is doubled?

20.7 Two identical charges of 1.0×10^{-7} C are separated by 1.0 cm. What is the mutual force if the charges are increased by a factor of 10 and the separation is kept at 1.0 cm?

Answers

20.1 $^{35}_{17}Cl$

20.2 4.8×10^{-9} N, attractive

20.3 10.7 N right

20.4 38.3 N directed 45° above the horizontal

20.5 0.9 N

20.6 0.225 N

20.7 90 N

Problems

1. Two equal small balls are charged with +30 μC and −6 μC respectively. Find the mutual force between them when their centres are 6 cms. apart, before and after contact with each other.

2. Two small balls, each one gm. in mass, are suspended from the same point by silk fibres 0.49 m. long. Each ball has a charge q. Find q such that the balls will separate to a distance of one cm.

3. Two small spheres 10 cm apart are charged with + 5 μC units and −5 μC units respectively. Find the direction and magnitude of the force acting on a +1 μC unit at a distance of 10 cms. from both charges

21

ELECTRIC FIELD

When two charges exert forces on each other, they do so through an invisible intermediate "field." Electric or **E** fields are the vehicles of electric force. Instead of having charges interact somewhat magically through empty space $(q \leftrightarrow q)$, each charge has an associated field that extends through space and communicates with other charges $\left(q \leftarrow \mathbf{E} \rightarrow q\right)$

E fields are defined in terms of the force **F** that they exert on a "test charge" q_0 that is placed in the field so that

$$\mathbf{E} = \mathbf{F}/q_0.$$

Note that this shows that the units for the field are N/C. It is usually said that the test charge is infinitesimally small so that it does not disturb the source of the **E** field. However, it is sufficient to know that when a charge Q is placed in an existing **E** field, it is subject to a force given by

$$\mathbf{F} = Q\mathbf{E}$$

21.1 An **E** field at a point P is given by $E_x=0$, $E_y= -34\times10^4$ N/C. Find the force exerted on a charge of 1.0×10^{-5} C when it is placed at point P. What is the force on a charge -1.0×10^{-5} C at the same point?

Point Charges

A point charge q or "point source" has an associated **E** field given by

$$\mathbf{E} = k\frac{q}{r^2}\hat{\mathbf{r}}$$

where the field point and source point are depicted in the figure.

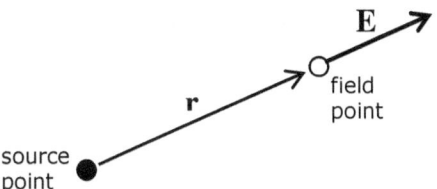

The direction of the electric vector is the same as the direction of the force on a *positive* test charge placed at the "field point." (That is, at a point a distance r from the source.)

The total **E** field at a given point is the vector sum of **E** fields due to all point sources. This is virtually the same process as vector summing forces on a particle.

21.2 Charges of 2×10^{-4} C are located at the two lower vertices of the triangle.
(a) Find the **E** field at the third vertex, point P.
(b) Find the force on a positive charge 1.0×10^{-5} C placed at P.

Continuous Charge Distributions *(Calculus Based)*

Electric fields due to continuous charge distributions are calculated by treating the distribution as a collection of infinitesimal point sources. The electric field for a linear charge distribution λ of length ℓ, for example, is given by

$$E = k\int \frac{\lambda\, d\ell}{r^2}\hat{\mathbf{r}}$$

21.3   Find the electric field at a distance L above the center of a flat circular wire of radius R that carries a uniform linear charge density λ.

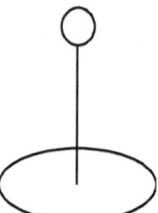

Practice Problem for Electric Fields

The following four multiple-choice questions pertain to this practice problem.

21.4   An electron is projected with an initial speed of 7×10^6 m/s into a uniform field between parallel plates. The direction of the field is vertically downward, and

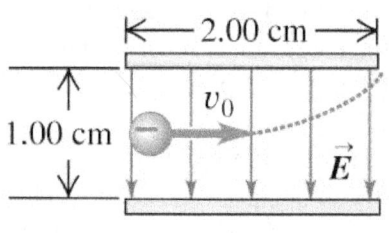

the field exists only between the plates. The electron enters the field at a point midway between the plates. The electron just misses the upper plate as it emerges from the field.

21 ELECTRIC FIELD

1. How long does it take for the electron to travel the horizontal length of the plates?
   (a) 2.86×10^{-9} s
   (b) 3.76×10^{-9} s, (c) 1.4×10^{-9} s,
   (d) 3.86×10^{-8} s, (e) 1.44×10^{-8} s

2. The electron has charge -e. The magnitude of the force on the electron is therefore
   (a) eE, (b) E/e, (c) 0, (d) e/E

3. Calculate the acceleration a required to meet the problem conditions in m/s^2.
   (a) 2.35×10^{14} (b) 1.22×10^{14} (c) 8.55×10^{13}
   (d) 3.12×10^{13} (e) $(1.9 \times 10^{13}$

4. Calculate the magnitude of the electric field in N/C
   (a) 8.85×10^3, (b) 8.00×10^2, (c) 9.17×10^3
   (d) 7.96×10^3, (e) 6.97×10^3

An Approximation

An approximation that is often useful, as in a dipole problem to follow, is

$$(1+x)^n \approx 1 + nx \quad \text{for small } x$$

21.5 Use the expression above to approximate the following without using a calculator:
   (a) $\sqrt{(1.04)}$, (b) $1/(.96)$, (c) $\sqrt{(4.16)}$, $1/(r-d)^2$ where $r \gg d$

Dipoles

When charges +q and -q are placed a small distance d apart, the system is called a *dipole*. Although a dipole has no net charge, it still produces a nonzero field because of the small separation between the charges of opposite sign.

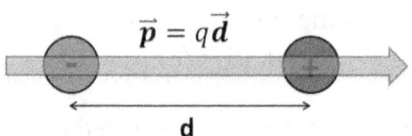

$$\vec{p} = q\vec{d}$$

d

Dipoles aquire potential energy U when they are placed in an electric field:

$$U = -\mathbf{p} \cdot \mathbf{E}$$

where the *dipole moment* **p** is given by

$$\mathbf{p} = q\mathbf{d} \qquad \mathbf{d} \text{ is directed from } -q \text{ to } +q$$

21.6   Show that for a point at distance r from the center of a dipole along its axis, that $E = 2k\,p/r^3$, when r is large relative to the dipole. Find the direction of **E**. (Use $(1+x)^n \approx 1+nx$ for small x)

Answers

21.1 $F_x = 0, F_y = -3.4\text{N}$ and $F_x = 0, F_y = 3.4\text{N}$

21.2 0.617×10^5 N/C directed straight up the page, 0.617 N

21.3 $E_y = k2\pi R\lambda L / \left(L^2 + R^2 \right)^{\frac{3}{2}}$

21.4   a, a, b, e

21.5   1.02, 1.04, 2.04, $1/r^2 + 2d/r^3$

21.6   Direction along dipole from –q to +q

134

Problems

1.   What are the magnitude and direction of the electric field
        1.5 m away from a positive charge of 2.1×10^{-9} C?

2.   Two point charges a and b both +20nC are placed 10m
        apart from one another.

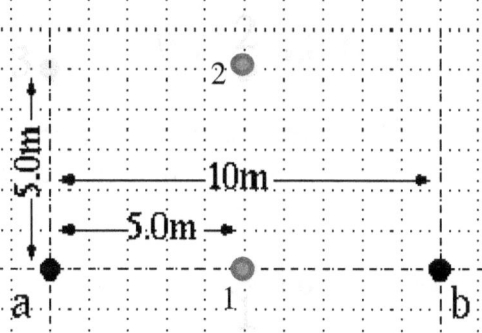

Find the electric field at point 1 centered between a & b.
Find the electric field at point 2 located 5.0m above the
center of the line joining a & b.

$$22$$

GAUSS'S LAW

Gauss's law is a mathematical restatement of Coulomb's law. It is very useful for finding the E fields of symmetrical charge distributions.

Field Lines and Gauss's Law

The electric field due to a point charge can be thought of as an aray of vectors, one at every field point, diminishing in size as we get further from the charge. The electric field vectors are pictured as originating on positive charges and terminating on negative charges. A useful representation results when we connect these 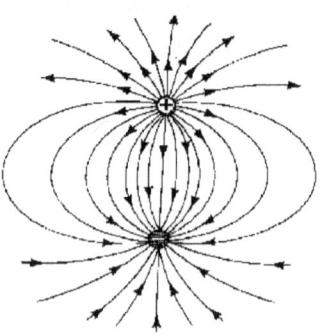 vectors into continuous lines. We can let the density of lines represent the field strength E.

We will soon find it convenient to write k as

$$k = \frac{1}{4\pi\varepsilon_0}$$

where ε_0 is called the *permittivity of free space*,

$$\varepsilon_0 = 8.85 \times 10^{-12} \; C^2/N\text{-}m^2$$

Calculus based physics texts show that Coulomb's law can be expressed in a mathematical form called *Gauss' law*. In the simplest symmetrical cases a charge q is enclosed in a specific closed area A and the electric field **E** is perpendicular to the surface of this area. In those cases Gauss' law can be written simply as

$$EA = \frac{q}{\varepsilon_0}$$

Charges outside of this closed area do not contribute to the integral. Note that if **E** is considered to be the density of field lines per unit area, then q/ε_0 is the number of field lines originating in the area. This is one of the four Maxwell equations governing all electric and magnetic fields; it expresses the Coulomb law in terms of the electric field.

Electric Flux

We visualize the E field as vectors (field lines) in space; the more dense the lines, the stronger the field. Let E represent the number of (imagined) field lines per unit area in the direction of **E**. The *electric flux* $d\phi$ through an element of area dA is then the number of lines through that area.

Integrating, we get an expression for flux through a finite area:

$$\varphi = \int \mathbf{E} \cdot dA$$

or in other notation

$$\varphi = \sum E_\perp \Delta A$$

where the integration or sum is understood to be over the entire surface. We can think of the flux as the number of field lines crossing the surface.

Gauss's Law

Gauss's law expresses the fact that the number of field lines is proportional to the charge q enclosed in a surface,

$$\varphi = \frac{q}{\varepsilon_0}$$

or, the equivalents

$$\int \mathbf{E} \cdot dA = \frac{q}{\varepsilon_0}$$

$$\sum E_\perp \Delta A = \frac{q}{\varepsilon_0}$$

Consider the special case of a point charge where we want to find the electric field at a distance r. Draw a spherical surface with charge q at its center. This useful sphere exists only in imagination–it is called a *Gaussian surface*. By convention, area vectors are perpendicular to the plane of the surface so the electric field **E** is everywhere perpendicular to the surface of area dA. Flux is then simply EA and the law is

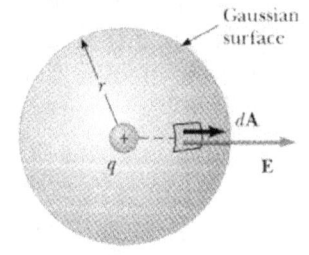

$$EA = \frac{q}{\varepsilon_0}$$

Substitute the area of the sphere, $4\pi r^2$ and solve for E. Identify $k = 1/4\pi\varepsilon_0$ to complete the familiar expression.

22.1   Show that applying Gauss's law to find the **E** field due to a point charge gives the Coulomb law.

Here is an intuitive argument (a so-called "handwaving" argument) to establish Gauss's law without a rigorous derivation: Gauss's law expresses the number of field lines exiting from charges enclosed within an arbitrary closed surface. The law reproduced the Coulomb law for a point charge. The number of exiting field lines does not change if we distort the spherical surface used in that demonstration, so we expect that the law must hold for arbitrary closed surfaces (Gaussian surfaces). Moreover, since charges and their associated field lines are additive, the law must also hold for arbitrary charge distributions within the Gaussian surface.

Applications

Note: Conductors have no charge in their interiors (that is, all charge on a static conductor is on the surface). This is a useful fact and is important for many problems.

22.2   Find the E field inside and outside a solid conducting sphere of radius R carrying total charge q.

22.3   Challenging problem: Find the E field (as a function of radial distance r) inside and outside a solid sphere of radius R with total charge Q uniformly distributed throughout. (This is *not* a conductor.)

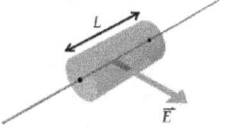

22.4   Find the E field outside of a long, straight line of charge with constant charge per unit length λ.

Here a Gaussian cylinder has the appropriate symmetry. The radial field lines are in the direction of the surface area so you can use $EA = q/\varepsilon_0$ with $q = \lambda L$ and $A = 2\pi r L$. Note that no field lines exit from the ends of the cylinder.

22.5   Show that the E field immediately above the surface of a conductor with surface charge density σ (charge per unit area) is given by

$$E = \sigma/\varepsilon_0$$

This result will be used when treating capacitors.

The appropriate Gaussian surface in this case is a pillbox shape. The top of the pillbox has area A so the enclosed charge is σA. The bottom of the pillbox is embedded in the conductor so no field contributes to the bottom face.

22.6   Show that the E field immediately above the surface of a charged sheet (*not* a conductor) with surface charge density σ is $E = \sigma/2\varepsilon_0$.

A Gaussian pillbox is appropriate, but here the lower face experiences a field contributing an additional EA to the left side of $\sum E_\perp \Delta A = q/\varepsilon_0$.

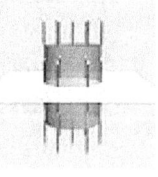

The following problem makes full use of the facts that (1) Gauss's law applies to total charge *inside* the imagined Gaussian sphere and (2) the field inside a conductor is zero.

22. 7   Consider two solid concentric conducting spherical shells. A +2C
charge is placed on the center conductor **A** and the outer shell **B** is
given a total charge +4C.
What is the total charge on the
i) inner surface of **A**;
ii) outer surface of **A**;
iii) inner surface of shell **B**;
iv) outer surface of shell **B**?

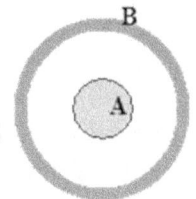

Answers

22.2   0 (**E**=0 inside conductor), kq/R^2 (like a point
charge)

22.3   Inside $r\rho/3\varepsilon_0$ with charge density $\rho = Q/\left(\frac{4}{3}\pi R^3\right)$,
outside kq/r^2

22.4 $\lambda/(2\pi\varepsilon_0 r)$

22.7   0, +2C, -2C (a sphere within **B** must have 0 field and
then contain 0 total charge), +6C (so the sum of inner
and outer charges totals +4C).

Problems

The next five multiple choice questions pertain to the following:
A conducting spherical shell B surrounds a concentric solid spherical conductor A as shown. A charge Q is placed on the solid sphere A and no charge is placed on shell B.

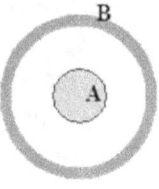

1. The electric field at a radial distance r inside the solid conductor A is
   (a) 0, (b) kQ/r^2 , (c) kQ^2/r^2 , (d) kQ^2/r, (e) kQ/r

2. The electric field at a radial distance r in the space between the conductors A and B is
   (a) 0, (b) kQ/r^2 , (c) kQ^2/r^2 , (d) kQ^2/r, (e) kQ/r

3. The charge on the surface of the solid conductor A is
   (a) 0, (b) Q, (c) -Q, (d) $2Q$, (e) -$2Q$

4. The charge on the inner surface of the spherical shell B is
   (a) 0, (b) Q, (c) -Q, (d) $2Q$, (e) -$2Q$

5. The charge on the outer surface of the spherical shell B is
   (a) 0, (b) Q, (c) -Q, (d) $2Q$, (e) -$2Q$

22 GAUSS'S LAW

23

ELECTRIC POTENTIAL

Electric field **E** is a vector property of the space surrounding a charge. Potential V is a scalar property of the same space. Of course, **E** and V are closely related and it is sometimes a choice of convenience as to which is used. **E** is intimately connected to the force **F**=q**E** whereas V is directly related to the potential energy $U=qV$. (Note that the words "potential" and "potential energy" do not mean the same thing.

Electric Potential

When a charge q does work W to move in and electric field from point A to B, the potential difference is defined as the negative of work done by the charge, -W per unit charge.

$$V_B - V_A = -\frac{W}{q}$$

A negative sign is included in the definition so that positive charges go spontaneously from higher to lower potentials, much like a ball rolls downhill from higher to lower potential energy.

The units of the V's are seen to be J/C. This unit is used so often that it has its own name, the familiar *volt* (written simply as "V").

23.1   A +2 C charge moves 0.1 m between two charged parallel plates   against an opposing force of 30 N so the work done by the charge is –30 X 0.1 J.   Find the potential difference from start to finish.

It is convenient to speak of a potential at a point, rather than potential differences. This is usually done by defining the point A in Eq. (**Error! Reference source not found.**) to be infinitely distant from the system of interest and assigning zero potential to this point; V_A=0. Then we are only concerned with the potential V_B and we drop the subscript.

$$V = -\frac{W}{q}$$

Now when we want to find a potential difference V_x-V_y, we can use this equation twice, once for each point x and y, and then subtract.

One of the most important properties of a potential difference V_B-V_A is that its value is the same regardless of the path the charge may take between A and B.

23.2   A +2 C charge expends –10 J to reach point A and –13 J to reach point B.   Find the potential difference from A to B.

Relation Between E and V

Potentials are scalar fields in space and can be depicted by *equipotential surfaces* as shown in the diagram by solid curves. These diagrams are like contour maps where the highest potentials are the highest points and positive charges move spontaneously to lower potentials much as a ball rolls downhill. Forces and electric fields are depicted by dotted lines and are necessarily perpendicular to the equipotential surfaces.

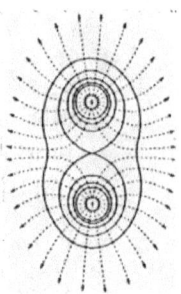

The link between E and V is given by

$$E = -\frac{\Delta V}{\Delta r}$$

or the calculus version,

$$E = -\frac{dV}{dr}$$

where **r** and **E** are perpendicular to the equipotential. surface.

The potential at a distance r from a point charge q is

$$V = k\frac{q}{r}$$

This should be memorized.

23.3   Calculus problem: Show that the potential $V = kq/r$ implies the correct electric field for a point particle.

23.4   Calculus problem: Use the integral form of $E = -dV/dr$ and the expression for the electric field of a point charge to derive $V = kq/r$.

Electric Potential Energy *U*

A charge q in an electrical potential V has potential energy U given by

$$U = qV$$

23.5   Find the potential energy stored in a charge $+2.0 \times 10^{-5}$ C when it is placed 3 m from a stationary charge of 10^{-5} C.

The following problem is simply treated using conservation of energy.

23.6   A movable $+2.0 \times 10^{-5}$ C charge has a mass of 1.2 kg. It is placed 3 m from a stationary charge of 10^{-5} C and is released from rest. What is its speed at a great distance from the stationary charge (escape velocity)?

Imagining a configuration to be assembled one charge at a time readily treats problems like the following. The first charge takes no energy to position, the second has potential energy due to the first, the third has potential energies due to each of the first two, etc.

23.7   Find the total energy stored in a configuration of three equal charges Q located at the corners of an equilateral triangle with side a.

Answers

23.1	1.5 V
23.2	1.5 V
23.5	0.6 J
23.6	1.0 m/s
23.7	$3kQ^2/a$

Problems

1. Positive charges of 1.5C, 4.2C, and 3.9C are placed at
    three corners A, B, C, of a square 30 cm on a side.
    Calculate the potential at the fourth corner D.

2. A sphere 10 cms. in diameter is. charged with +0.5C
    Find the potential at the surface of the sphere and
    at a point 20 cm. distant from its surface.

3. Find the total energy stored in a configuration of four
    equal charges Q located at the corners of a square with
    sides a.

23 ELECTRIC POTENTIAL

24

CAPACITORS

Capacitors are devices that store electrical energy by keeping positive and negative charges separated. Unlike batteries, they can discharge their energy in a sudden burst as can be seen in the discharge of a flashbulb or the shock of a cattle prod.

Capacitance

Two separate conductors having a potential difference applied across them can acquire opposite charges of equal magnitude. The device is called a *capacitor* and is denoted in circuit diagrams by the symbol —||—

The defining equation for capacitance C is

$$C = Q / V$$

where Q is the magnitude of charge stored on either conductor and V is the potential difference needed to apply the charge. The units of C (coul/volt) are called *farads;* f. Most often, we deal with microfarads, $1 \mu f = 10^{-6} f$.

24.1   A potential difference of 500 volts is applied across a $3\mu f$ capacitor. Find the charge on each plate.

The capacitance of a parallel plate capacitor is given by

$$C = K\varepsilon_0 A / d$$

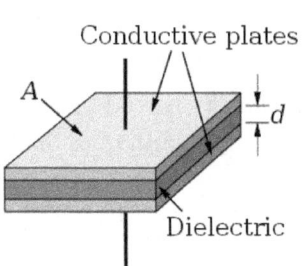

Conductive plates

Dielectric

where K is called the *dielectric constant* ($K = 1$ in vacuum), A is the plate area, and d is the distance between plates.

The energy stored in the electric field between plates is

$$U = CV^2 / 2$$

24.2   Use $C = Q/V$ and $U = CV^2/2$ to show the vacuum energy per unit volume in the electric field is $\frac{1}{2}\varepsilon_0 E^2$. This result applies generally and reveals that energy stored in an E-field is proportional to E^2.

24.3   Show that U has alternative forms
$$U = QV/2 \text{ and } U = Q^2/2C$$

The latter expression is readily derived with calculus. A change in energy dU due to the addition of charge dQ is $dU = VdQ$. The result follows by substituting $V = Q/C$ and integrating. All the other forms for U can then be found.

Capacitors in Series

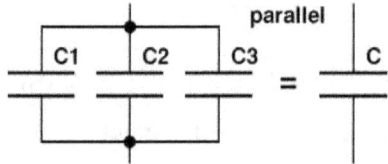

Capacitors linked in series as in the figure above can be replaced by a single equivalent capacitor C given by

$$\frac{1}{C} = \frac{1}{C_1} + \frac{1}{C_2} + \cdots \qquad (2)$$

Capacitors in Parallel

Capacitors in parallel as shown above can be replaced by an equivalent capacitor given by the sum of individual capacitors,

$$C = C_1 + C_2 + \cdots \qquad (3)$$

24.4   Find the equivalent capacitance of the circuit shown in the figure. All capacitance is given in microfarads.

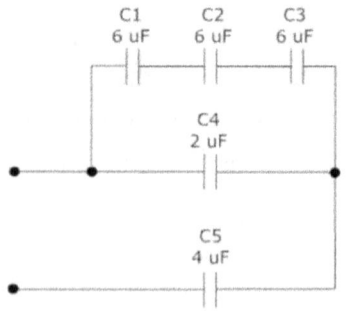

The next exercise makes use of the fact that a change in potential V depends only on initial and final points and not on the paths between. The figure below indicates that the potentials across capacitors C_1 and C_2 are equal. Then repeated use of the defining relation $CV = Q$ solves the problem.

24.5   Given $C_1 = 6\mu f$ and $C_2 = 3\mu f$ for the two parallel capacitors depicted in the diagram, find the charge across each under an 18V potential. Check that the sum of these charges equals the charge on the equivalent single capacitor.

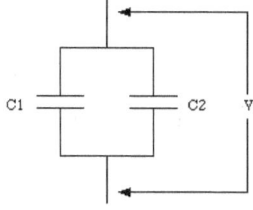

In the case of capacitors in series, it is important to realize that the charge stored in each capacitor is the same. This can be seen by considering the "internal" plates: *i.e.*, the charges on the negative plate of capacitor 1, and the positive plate of capacitor 2 must add to zero as is the case when no voltage is applied.

24.6   Given $C_1 = 6\mu f$ and $C_2 = 3\mu f$ for two capacitors connected in series. Find the charge across each under an 18V potential and find the potential difference across each. Check that the sum of these potentials equals the overall 18V potential.

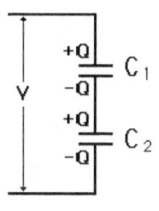

Answers

24.1	1.5×10^{-3} C and -1.5×10^{-3} C
24.4	2 µf
24.5	0.108mC, 0.054mC
24.6	0.036mC each, $V_1 = 6V$, $V_2 = 12V$

Problems

1. Find the equivalent capacitance of the circuit shown in the figure. All capacitance is given in microfarads.

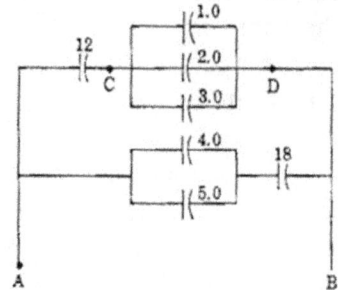

2. A potential of 18V is applied across the circuit shown in the figure. Find the potentials and charges across each capacitor.

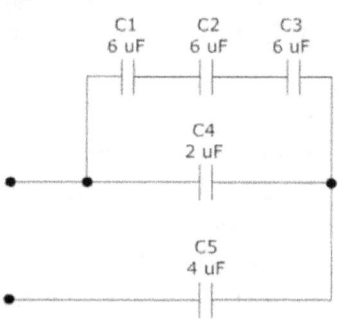

24 CAPACITORS

25

CURRENT & RESISTANCE

Charges flow through wires much like water flows through pipes. The flow of charge is called *current*. Potential differences supply the energy to drive the flow against the resistances inherent in the wires.

Current

Current I is a measure of the rate of transfer of charge. It is defined by the expressions

$$I = \Delta q/\Delta t \quad \text{or} \quad I = dq/dt \,,$$

and the units (C/s) are called *Amperes*, A.

A sign convention:
Current is positive (negative) when it describes the flow of positive (negative) charge. However, the current for the flow of negative charge in one direction is equivalent to the flow of positive charge in the opposite direction.

25 CURRENT & RESISTANCE

25.1   Electrons have a charge magnitude of 1.6×10^{-19} C. Given that 5×10^{20} electrons pass through a wire in 1.0 min, what is the magnitude of the current through the wire?

Current density **j** is a vector in the direction of current flow and describes the current flow per unit area. Its magnitude is given by
$$j = I/A, \qquad\qquad)$$
where A is the area through which current I flows.

Ohm's Law

Current in a segment of wire is proportional to the potential difference V across the wire. This is true for many materials and conditions. The phenomenological law, Ohm's law, is written as
$$V = IR,$$
where the proportionality constant R is a property of the conductor called *resistance*. The unit of resistance (V/A) is ohms, Ω. Resistance in a circuit is depicted by the symbol

When the potential difference V is supplied by a device like a battery or generator, it is referred to as a source of *emf*, electromotive force, and often written as ε. In a circuit, an emf is often drawn as

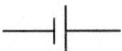

where the long side represents the positive terminal.
An altenative form is a circle as shown here:

The emf boosts positive charge to a higher potential. Emf is an idealized potential difference in that it is resistance-free (it can be measured in an open circuit without a current flow).

25.2 Electrons have a charge magnitude of 1.6×10^{-19} C. Given that 5×10^{20} electrons pass through a 20 Ω resistor in 1.0 min, find the potential difference across the resistor.

25.3 Find the current I for the circuit in the diagram given $\varepsilon = 18$ V and $R = 6$ Ω.

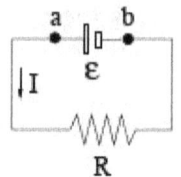

Resistance

Resistance is defined by the Ohm's law expression $R=V/I$. The resistance of a particular conductor depends upon its material, its size and shape, and its temperature. For a wire at a given temperature, resistance is proportional to the length of the wire L and inversely proportional to the cross sectional area A:

$$R = \rho\, L/A.,$$

The proportionality constant ρ (rho) is called *resistivity*. (Its defining equation is $\rho=E/j$. where E is the electric field.)

25.4 A 4 ohm length of wire is cut in half and the two lengths are placed together. What is the resistance of the composite?

Electrical Power

Recall that power P is defined by the rate of change of work, $P = \Delta W/\Delta t$ or $P = dW/dt$, and its units (J/s) are Watts, W. The power expended by a current is given by any of the following:

$$P = IV$$
$$P = I^2 R$$
$$P = V^2/R$$

25 CURRENT & RESISTANCE

Each form is obtained by substitution from $V = IR$ into $P = IV$.

25.5   A 100 W light bulb is lit by a 120 V generator. Find the resistance of the filament.

25.6   Use $V = IR$ and $P = IV$ to derive $P = I^2R$ and $P = V^2/R$.

25.7   Find the power expended by the circuit of problem 25.3.

Answers

25.1	1.33 A
25.2	27 V
25.3	3 A
25.4	1 ohm
25.5	144 ohm
25.7	54 W

Problems

1.   A simple circuit consists of an emf of 10V and total resistance $1k\Omega$. Calculate the power expenditure.

2.   A simple circuit consists of an emf of 18V and current 9mA. Calculate the power expenditure.

3.   A simple circuit carries current 10mA and total resistance $1k\Omega$. Calculate the power expenditure.

4.   The circuit in the diagram is known to have current 20 mA running throughout the loop. Find the power expended by each resistor.
Use these results to find the potential drop across each resistor.

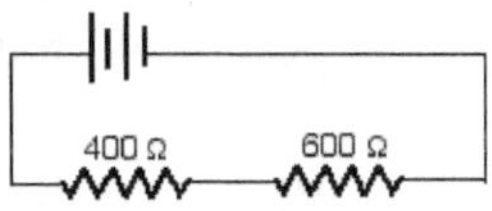

26

DC CIRCUITS

The currents in direct current (DC) circuits are assumed not to change in time. We say they are *steady-state* currents. Perhaps the most fundamental problem in DC circuits is to find the currents in all branches, given the emf's and resistances.

Potential Drops

In analyzing DC circuits, we follow branches of the circuit in a particular direction (usually arbitrary). As we encounter resistors and emfs, the following rules apply:

(a)   When the chosen path moves through a resistor R in the direction of the current, the potential change is $-iR$. This is spoken of as a "potential drop." When the path direction opposes the current, the potential change is $+ iR$

(b)   When the chosen path moves through an emf ε in the positive direction, the potential change is $+\varepsilon$. When the path direction is opposite, the potential change is $-\varepsilon$.

One of the most useful facts in analyzing circuits is that the potential drop between two points is the same regardless of the path taken between them.

26.1   The circuit carries a current of 2 A. Find the potential difference between points a and b, Vab. Do the calculation twice by going clockwise, and then counterclockwise around the circuit (a consistent result will assure that you are correctly applying the rules).

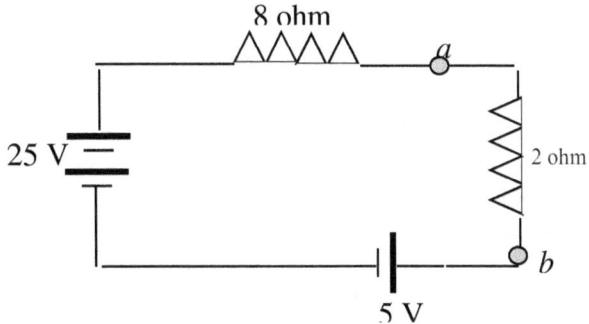

26.2   Given that $I = 1.5$ A in the circuit segment shown, find the potential difference $V_{ab} = V_b - V_a$.

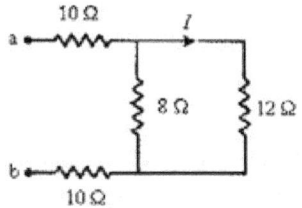

26.3   Given that $I = 2$ A in the circuit segment shown,
(a) find the potential difference across the 12 Ω resistor. (b) Use the result to find the current across the 8 Ω resistor.

26.4 Given that the current through the battery is 2 A, (a) find the potential difference across the 6 Ω resistor. (Note that the potential difference across the left branch must be the same as that across the middle branch because they both have the same "end" points.) (b) Use the result to find the current in the 3 Ω resistor.

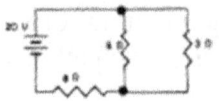

Series and Parallel

Resistors in series as shown may be replaced by a single equivalent resistor R as the sum of the series:

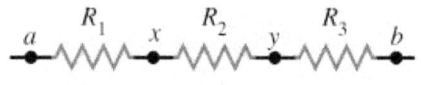

$$R = R_1 + R_2 + \cdots \quad \text{(series)}$$

Resistors in parallel may also be replaced by a single equivalent resistor R. The expression for parallel resistors is given below

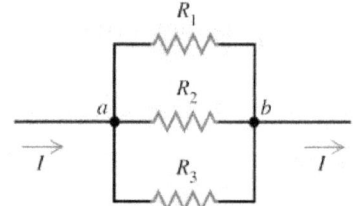

$$\frac{1}{R} = \frac{1}{R_1} + \frac{1}{R_2} + \cdots$$

Note: Resistors are *not* in series if a wire branches off from a wire segment between the resistors.

Note Resistor are *not* in parallel if an emf interrupts the connections.

26.5 Find the currents in all branches of the circuit. (Reduce the parallel and series resistors until only one resistor remains. Then apply Ohm's law to the simpler loop to find the current through the emf. Then use potential differences as in problem 25.4)

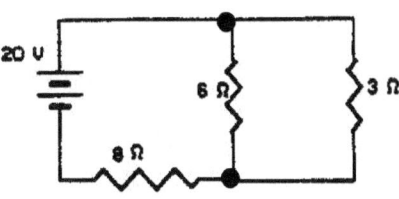

Kirchhoff Rules

When circuits cannot be reduced using series and parallel expressions, we use the powerful but laborious Kirchhoff loop and point rules:

i) **Point rule**: The sum of all currents into a branch point equals the sum of all currents going out of a branch point. If there are n branch points, only $n-1$ are independent.

ii) **Loop rule**: The sum of all potential changes around a closed loop is zero.

Apply these rules to different points and loops (pick the most convenient ones) until you have enough equations for unknowns. (Exhaust the points first because they are easier to handle but don't include redundant points that express the same equation.)

26.6 Find the currents in all branches of the circuit below. (This will require solving a set of simultaneous linear equations—typical for Kirchhoff problems.)

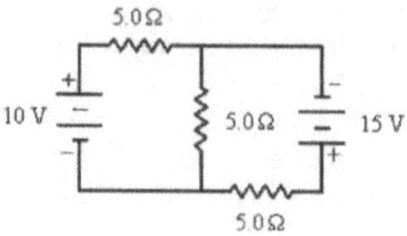

Varied Practice Problems

26.7 Given that ε = 30 V, find the current in all branches of the circuit shown.

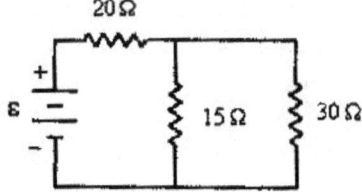

26.8 Given that ε = 8 V and R = 2 Ω, calculate the rate at which thermal energy is generated in the diagonal resistor shown in the diagram.

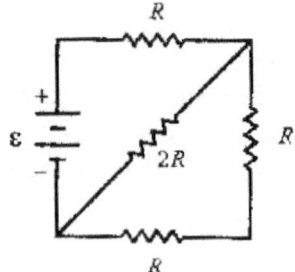

26.9 Given that the current in the left emf is 7/3 A, find currents in the other branches. (It is not neces- sary to know ε.)

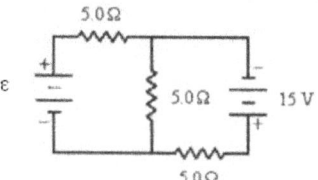

Mesh Analysis

When a circuit contains n independent loops, Kirchhoff's Volt- age Law, KVL, enables us to solve for n unknowns. You have solved problems with the "loop rule" before. Now we want to introduce a slight change in the approach in order to make it eas- ier and faster to apply.

Each loop in a circuit that does not contain another loop is called a *mesh*. The following is called mesh analysis. Each mesh is as- signed a loop current and a resistor that is adjacent to two loop currents, say i_1 and i_2, has a voltage drop of magnitude $(i_1 - i_2)R$.

example
A KVL analysis of the left mesh in the figure gives the follow- ing result:

$$V - iR_o - (i_1 - i_2)R = 0$$

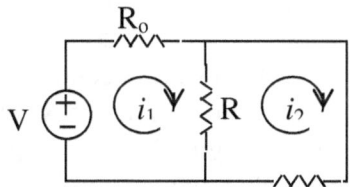

Of course, the current in the center segment is $(i_1 - i_2)$ down- ward.

26.10 Find the currents in all branches in the network using mesh analysis. The ground symbol indicates a zero voltage at that point. Calculate the voltage V_o common to the upper right segment from your results.

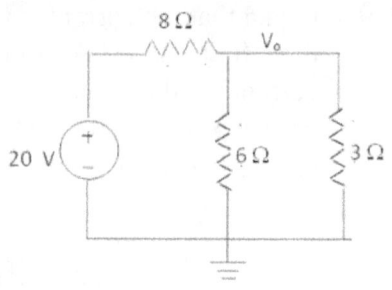

Notice that mesh analysis is more convenient for finding currents than for voltages.

Answers

26.1	−4 V
26.2	−22 V
26.3	−24 V, 3 A
26.4	4 V, 1.33 A
26.5	2 A through the emf, 2/3 A in the middle branch, 4/3 A in the right branch
26.6	7/3 A in the 10 V emf, 8/3 in the 15 V emf, 1/3 A upward in center branch
26.7	1 A in the left branch, 2/3 A in the middle branch, 1/3 A in the right branch
26.8	4 W
26.9	1/3 A upward in center branch, 8/3 A down through the 15 V emf
26.10	left to right: 2 A clockwise, 2/3 A down, 4/3 clockwise, V_0 = 4 V

26 DC CIRCUITS

Problems

1.   Find the equivalent resistance of the circuit below and calculate the current through the emf source.

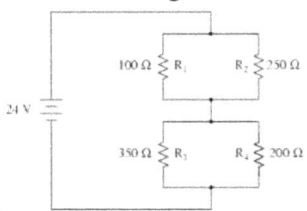

2.   Calculate the currents through each resistor and the potential drops across each.

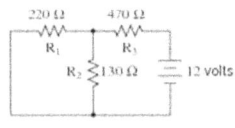

3.   Calculate the currents through each resistor and the potential drop across each.

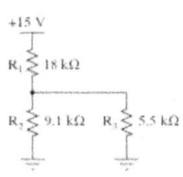

4.   Calculate the currents through each resistor and the potential drops across each.

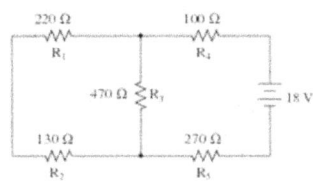

5.   Calculate currents $I_1, I_2,$ and I_3.

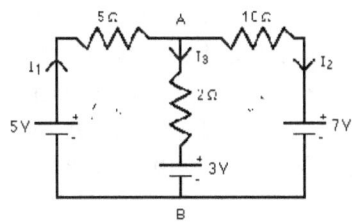

27

MAGNETIC FIELD

Moving charges or continuous currents generate magnetic fields. These vector fields exert forces on other moving charges or currents. This section describes the magnitude and direction of the magnetic force (it is left to the next section to describe the production of the magnetic field).

The Magnetic Field

The magnetic field or **B** field is defined by its effects on moving charges or currents. . The unit for B (N-s/C-m) is called a *tesla*. A charge q moving with velocity **v** in a magnetic field **B** experiences a force **F** that is perpendicular to both **v** and **B** as shown in the fig-

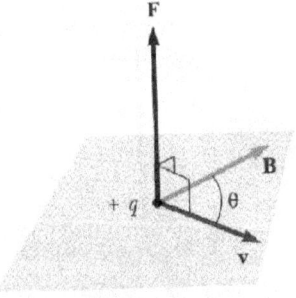

ure. Moreover, the force has a magnitude given by

$$F = qvB \sin \theta$$

where θ is the angle between **v** and **B** and the direction is perpendicular to the plane containing **v** and **B** as seen in the figure.

A mathematical form of this same force law is summarized by a *cross product*:

$$\mathbf{F} = q\,\mathbf{v} \times \mathbf{B}$$

The direction of this force is given by a *right-hand-rule* for magnetic force as shown in the figure. Again, the magnitude of **F** is $F = qvB \sin \theta$.

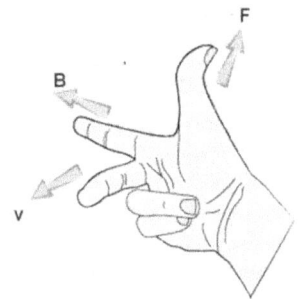

27.1   A charge of 1.6×10^{-19} C moves to the right in the plane of the page at velocity 0.5×10^8 m/s. A magnetic field of 2 tesla is directed into the page. Find the force (magnitude and direction) on the charge.

27.2   A charge of -1.6×10^{-19} C moves to the right in the plane of the page at velocity of 0.5×10^8 m/s. A 2 tesla magnetic field is in the plane of the page at an angle of 30° above the direction of motion. Find the force (magnitude and direction) on the charge.

An **E** field exerts a force $q\mathbf{E}$ on a charge, and the combined force due to this and the **B** field is called the Lorentz force:

$$\mathbf{F} = q\mathbf{E} + q\,\mathbf{v} \times \mathbf{B}$$

Force on a Current

Charges in motion comprise a current. This is expressed in the relation,

$$q\,\mathbf{v} = i\,\mathbf{L}$$

where **L** Is a directed length of wire. The form of $\mathbf{F} = q\,\mathbf{v} \times \mathbf{B}$ becomes

$$\mathbf{F} = i\,\mathbf{L} \times \mathbf{B}$$

27.3 A square circuit with sides of length L carries current i. The circuit is placed with a pair of opposite sides parallel to a uniform magnetic field **B**. Calculate the magnitude of torque τ on the circuit.

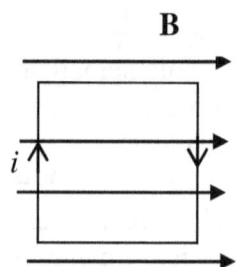

Problem 27.3 illustrates a special case of torque on a current carrying loop:

$$\tau = \mu \times B$$

where the *magnetic moment* μ is given by the product of the number of turns in the loop N, the loop current i and the directed loop area **A**,

$$\mu = N i \, A$$

The magnetic potential energy U stored in a magnetic dipole is

$$U = -\mu \cdot B$$

Circular Motion of a Charge In a B field

A charge moving perpendicular to a uniform B field moves in a circle. The magnetic force qvB is equal to the centripetal force mv^2/r,

$$qvB = \frac{mv^2}{r}$$

27.4 Protons are accelerated through a potential difference of 1000 volts and then enter a magnetic field of 0.1 tesla perpendicular to their direction of motion. Calculate the radius of their path. $m_p = 1.67 \times 10^{-27}$ kg , $e = 1.6 \times 10^{-19}$ C.

Answers

27.1 $\quad 1.6 \times 10^{-11}$ tesla directed "up" parallel to the page

27.2 1.6×10^{-11} tesla directed "down" parallel to the page

Problems

1.   A uniform magnetic field with magnitude $B=1.2 \times 10^{-3}$ T points vertically upward throughout a chamber. An electron with a velocity $v = 3.2 \times 10^7$ m/s enters the chamber moving horizontally from south to north. Find (a) the magnitude and (b) the direction of magnetic force on the electron.

2.   A particle with charge $q = 10^{-5}$ C moves to the right with velocity $v = 10^2$ m/s in a magnetic field $B = 2$ N-s/C-m which makes an angle of $30°$ with **v**. Find the magnitude of the force on the charged particle.

3.   Two charged particles with the same mass and charge q move in circular paths in a magnetic field **B** that is perpendicular to their velocities. If particle A has a velocity v_A and particle B has a velocity v_B, compare the periods for the particles to make one complete orbit.

4.   Two smooth conducting rails a distance $L = 2.0$ m apart make an angle of $22.5°$ with the horizontal. A bar of mass 1.2-kg rests across the rails, as shown below. A uniform magnetic field 0.50 N-s/C-m is vertically upward. A battery is connected to cause a current I to flow through thebar as shown. (a) Draw a side view of the bar and indicate the direction of each force that acts on it neglecing friction. (b) What current I will allow the bar to remain at rest?

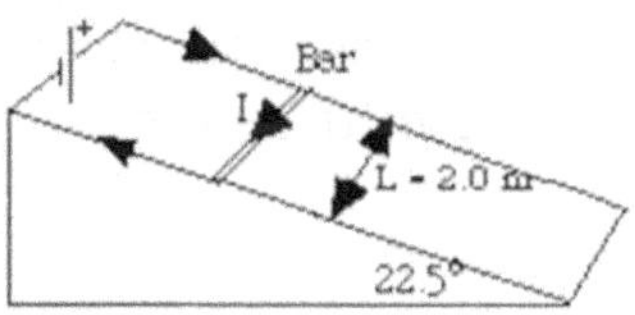

28

AMPERE'S LAW

The last chapter addressed the effects of a magnetic field B on a current. Here we discuss how a B field originates from a current.

A magnetic field is depicted by vectors in space. By convention, these are taken to originate from a permanent magnet's north pole and terminate at the south pole as shown. Recognize that while there are elementary charges, there are no corresponding elementary magnets. Magnetism is caused by moving charges or currents. Permanent magnets owe their magnetism to electrons orbiting atoms in the magnetic material.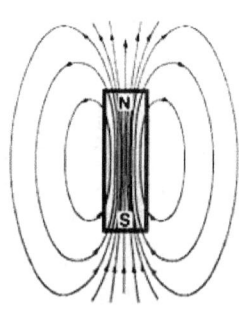

This chapter distinguishes between calculus-based and algebra-based treatments. As usual, algebra-based courses should simply skip items specifically labeled for calculus. Here the two tiered approach necessitates some repetition.

172

Blot-Savart Law (Calculus Based)

A segment of wire $d\mathbf{L}$ carrying current I produces an increment of magnetic field $d\mathbf{B}$ at the point shown. The contribution of the infinitesimal segment is given by

$$d\mathbf{B} = k_m I \frac{d\mathbf{L} \times \hat{\mathbf{r}}}{r^2}$$

where k_m is a universal constant,

$$k_m = \mu_0 / 4\pi = 10^{-7} \text{ T-m/A}$$

The direction of $d\mathbf{B}$ is the same as the direction of the cross product $d\mathbf{L} \times \mathbf{r}$. Vector $d\mathbf{L}$ is in the direction of the current and vector $\mathbf{r}$ goes from the source point to the field point as shown.

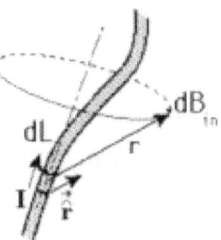

28.1 A circular loop carries current I as shown. Show that the magnitude of the $\mathbf{B}$ field at the center is

$$B = \frac{\mu_0 I}{2 r}$$

Confirm the direction of the magnetic field is as shown in the diagram.

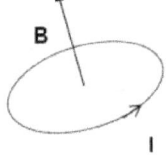

Ampere's Law (Calculus Form)

Calculus version: We found that for some problems it was more convenient to use the Coulomb law in an alternate mathematical form, Gauss's law. The Biot-Savart law is analogous to Coulomb's law and it too has another convenient form called Ampere's law. Instead of using a Gaussian surface, this law refers to an arbitrary closed loop called the Amperian circuit.

Ampere's law looks a lot like Gauss's law and is used in similar ways:

$$\oint_c \mathbf{B} \cdot d\mathbf{L} = \mu_0 I_{\text{enclosed}}$$

The integral is around the arbitrary Amperian circuit, and the current I is the current inside the circuit.

28.2 Calculus based: Show that the magnetic field at a distance r from a long, straight wire carrying current I is

$$B = \frac{\mu_0 I}{2\pi r}$$

A right-hand rule decides the direction of the field. Show this follows from Biot-Savart. Students usually memorize the result because it is treated as a "known" quantity in some problems.

Ampere Special Cases (Algebra and Calculus)

Instead of relying on the general form of Ampere's law, you can use the following important special cases to treat most problems. In these cases, μ_0 is a universal constant

$$\mu_0 = 4\pi \times 10^{-7} \text{ T-m/A}$$

- The magnetic field due to a long wire carrying current I at a distance r perpendicular to the wire,

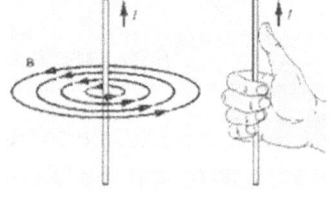

$$B = \frac{\mu_0 I}{2\pi r}$$

The direction of the field is given by the right-hand rule depicted in the figure above. That is, the thumb is directed along the current and the curl of the fingers then indicates the **B**-field direction.

- A circular loop carrying current I has a **B** field at the center with magnitude,

$$B = \frac{\mu_0 I}{2\,r}$$

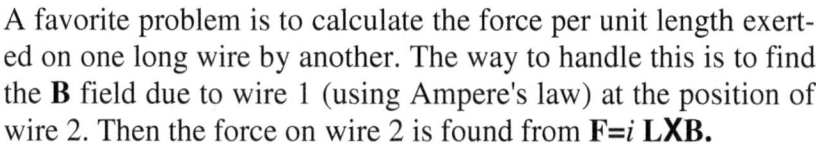

Again, the field direction is determined by the right hand rule. Some imagination is needed to see how the magnetic field represented by curled fingers leads to the appropriate direction at the center of the loop.

A favorite problem is to calculate the force per unit length exerted on one long wire by another. The way to handle this is to find the **B** field due to wire 1 (using Ampere's law) at the position of wire 2. Then the force on wire 2 is found from **F**=i **LXB.**

28.3 Two long, straight and parallel wires separated by a distance r carry current i in opposite directions. Find the magnitude and direction of the force per unit length on the wires. Variations of this problem include having different currents in each wire, having the currents in the same direction, and including additional wires.

- We often need to know the **B** field inside a solenoid (a wire wound closely in a helix around a cylindrical shape) or a toroid (a solenoid bent into a circular shape). The result is a special case for a solenoid or toriod with N turns per length L carrying current I is

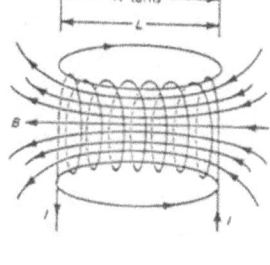

$$B = \mu_0 I \frac{N}{L}$$

28.4 Calculus based: Show that the B field inside a solenoid with N turns per length L carrying current I is

$$B = \mu_0 I \, N/L$$

When we make the approximation that the B field vanishes outside the coils, the computation of the interior **B** field is easily done using Ampere's law. Students often memorize this result because it is treated as a "known" quantity in some problems.

28.5   Classical physics determines that an electron orbiting in a circular orbit of radius r with tangential velocity v is equivalent to a current $I = ev / (2\pi r)$. Find the magnetic field magnitude at the center of this orbit. Is the field stronger or weaker with increasing r?

Answers

28.3 $\dfrac{F}{L} = \dfrac{\mu_0 i^2}{2\pi r}$ directed away from the wires

28.5 $B = \dfrac{\mu_0 ev}{4\pi r}$ weaker

Problems

1.   The figure below represents two wires carrying currents into the page. (a) What is the magnitude of the magnetic field at point P_1, which is halfway between the wires, if both wires carry current I. (b) What is the direction and magnitude of the magnetic field at P_2.? (c) If a proton moves out of the paper at P_2 parallel to the current-carrying wires, what is the direction of the force exerted on it?

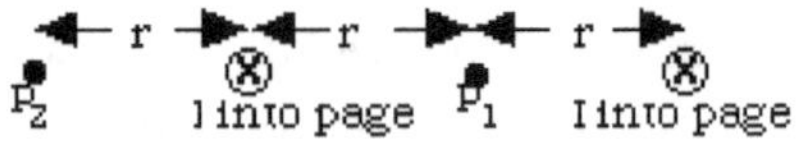

2.   Two very long wires carrying a current I of 3.0 A out of the page are at two of the corners of an equilateral triangle, as shown. The sides of the triangle are 1.0 m long. Find the direction and magnitude of the magnetic field at P, the third corner of the triangle.

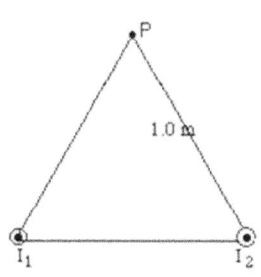

3.   A very long straight wire carries a current I = 10 A and the rectangular loop carries a current I′= 20 A as indicated in the figure. Find the magnitude and direction of the force on the loop. a = 0.01 m, b = 0.03 m, and L = 0.03 m.

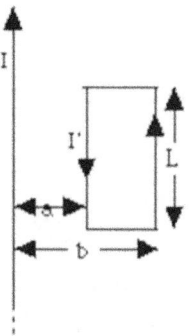

4.   In the coaxial arrangement shown in the figure, a current I in the inner cylinder of radius a is out of the page and a current I in the outer cylindrical shell of inner radius b is into the page. The outer shell of radius c also carries current I into the page. Find the magnetic field (direction and magnitude) for

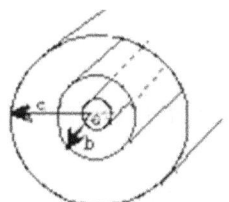

(a) $r \leq a$,
(b) $a \leq r \leq b$,
(c) $b \leq r \leq c$, and
(d) $r \geq c$.

29

FARADAY'S LAW

Moving a magnet toward a wire loop causes a surge of current in the loop. This is a special case of Faraday's law where a changing magnetic field creates a surrounding E-field. When a wire is introduced in this E-field, the wire carries a current as if a battery is driving it. We follow convention in this course by describing the E-field indirectly through the fictitious battery or emf ε.

Faraday's Law

Magnetic flux ϕ through a flat surface of area $\mathbf{A}$ is defined by

$$\varphi = \mathbf{B} \cdot \mathbf{A}$$

or, more generally, in calculus notation,

$$\varphi = \int \mathbf{B} \cdot d\mathbf{A}$$

The units of ϕ (tesla-m^2) are *webers*.

Faraday's law requires that a *changing* φ induces an emf around the perimeter of the surface area,

$$\varepsilon = -\frac{\Delta\varphi}{\Delta t} \quad \text{or} \quad \varepsilon = -\frac{d\varphi}{dt}$$

The negative sign indicates that the induced emf will oppose the change of flux, but we need not be concerned with it yet.

This figure shows a permanent magnet approaching a wire loop. This causes a flux change in the loop area inducing a current I to flow in the loop.

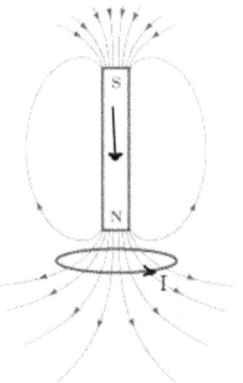

Often the wire loop has N wire turns. Since the emf is N times as great as it would be for one loop, the Faraday law is usually written as

$$\varepsilon = -N\frac{\Delta\varphi}{\Delta t} \quad \text{or} \quad \varepsilon = -N\frac{d\varphi}{dt}$$

29.1 A 100-turn coil loop of area 10^{-2} m^2 is perpendicular to a magnetic field of 0.3 tesla. The field is reduced to zero in 0.2 seconds. Find the average emf induced in the loop. The same result would obtain if the loop area was reduced to zero in the same time.

Motional Emf

When a wire is moved through a B-field, it traces out an area in a given time. That is, it causes a changing magnetic flux and an emf is created in the wire. A straight wire of length L moving at velocity v perpendicular to a uniform magnetic field **B** has an induced emf given by

$$\varepsilon = BLv$$

The following exercise asks to derive this from the Faraday law. Start with $|\varepsilon| = \Delta\phi/\Delta t$ and note that $\Delta\phi = B\,\Delta A = B\,L\Delta t$.

29.2 Derivation: Show that a straight wire of length L moving at velocity v perpendicular to a uniform magnetic field **B** has an induced emf given by $\varepsilon = BLv$.

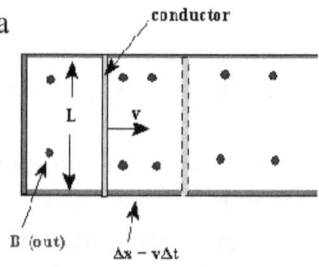

conductor

L

v

B (out)

Δx ~ vΔt

Lenz's Law

Induced current must flow in a direction that causes an associated induced B field to oppose the change in flux that created it. The term *induced current* pertains to the current due only to Faraday's law.

29.3 A magnet falls through a loop of wire as shown. Indicate the direction of induced current flow when
(a) the magnet approaches the plane of the wire loop,
(b) the magnet recedes from the plane of the wire loop.

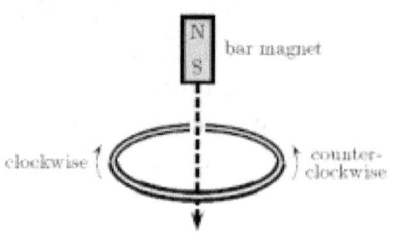

bar magnet

clockwise

counter-clockwise

The following five practice problems pertain to this system.

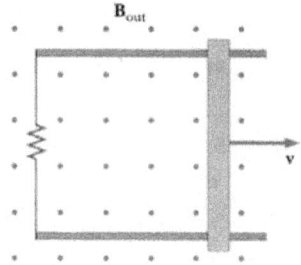

The rod in the diagram is moving at a speed of 5.0 m/s perpendicular to a magnetic field of 0.8 T. The rod has a length of 1.6 m and has negligible resistance as do the rails. The resistor is the filament of a light bulb with resistance 96 Ω

29.4   Find the motional emf across the rod.
         (a)1.0 V, (b)1.4 V, (c) 3.2 V, (d) 6.4 V, (e) 8.0 V

29.5   The current through the resistor (in amps) is
         (a) 0.067 A, (b) 0.56 A, (c) 1.0 A, (d) 1.44 A, (e) 2.3 A

29.6   What is the electrical power produced in the resistor?
         (a) 75 W, (b) 50 W, (c) 15 W, (d)1.7 W, (e) 0.43 W

29.7   What force is required to keep the bar moving?
         (a) 0.085 N, (b) 0.22 N, (c) 0.35 N, (d) 0.7 N, (e) 0.9 N

29.8   Which of the following pairs indicates the respective directions of current flow in the rails and the force of the field on the bar?
         (a) clockwise and left,
         (b) clockwise and right,
         (c) counterclockwise and left,
         (d) counterclockwise and right

Answers

29.1	1.5 V
29.3	clockwise, counterclockwise
29.4	d
29.5	a
29.6	c
29.7	a
29.8	a

Problems

1. A square loop of wire, 2.0 m on a side, has a resistance of 0.04 Ω. The loop is in a horizontal plane. Initially a magnetic field **B** = 0.080 T points vertically downward. If this magnetic field is reduced in magnitude at a constant rate to 0.040 T in 2.0 s, find the current and the sense of the current as you look down on the loop. Note: 1 T = 1 N/A-m.

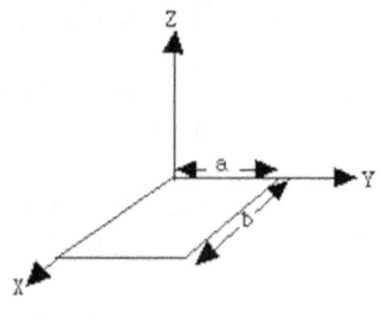

2. A bar of mass m = 0.50 kg slides downward along stationary wires that are separated by a distance L = 0.5 m and are joined at the top by a resistance R = 0.010 Ω. A uniform magnetic field B = 0.20 N/A-m points into the paper. (a) If the bar slides down, what is the sense of the current through it? (b) At what speed v will the bar experience no force?

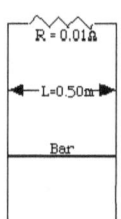

3. A rod with length ℓ, mass m and resistance R sl ides without friction down two parallel conducting rails inclined at an angle θ with the horizontal. The rails and the bar that connects them at the bottom of the incline have negligible resistance. Find the constant velocity v achieved by the wire in a uniform magnetic field **B** directed vertically upward.

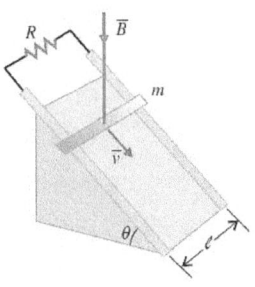

30

INDUCTANCE

According to the Faraday law, a current is induced in a wire loop when the loop surrounds a changing magnetic flux. This induced current has the property of opposing the flux that produced it. Systems that experience induced currents due to their own flux changes are said to have *self inductance* and systems with induced currents due to flux changes generated from outside experience *mutual inductance*.

Self- Inductance

A current flowing in a coil creates a magnetic field (by Ampere's law). If this current changes, the B field also changes and, by Faraday's law, a new (smaller) *induced* current arises in the coil. We have seen that this induced current opposes the original current by Lenz's law. This property of a coil (or other structure) "resisting" an original current is called self-inductance L. Its MKS units are *henries* and it is depicted in circuit diagrams with the symbol ⎓⎓⎓⎓⎓

The basic self-inductance relation is

$$\varepsilon = -L\,\Delta I/\Delta t \quad \text{or} \quad \varepsilon = -L\,dI/dt$$

30.1   A coil carries a current of 2 amp which is reduced to zero in 0.8 sec. This causes an induced emf of 0.5 volt. Find the self-inductance.

30.2   Derivation: Consider a solenoid of length ℓ and cross section area A with N windings. Recall that the B field in a solenoid is given by $B = \mu_0 I\,N\,/\,\ell$ and show that the self-inductance of a solenoid is $L = N^2 \mu_0 A/\ell$.

The energy stored in a self-inductor is the energy stored in the magnetic field:

$$U = \tfrac{1}{2}LI^2$$

30.3   A coil of 0.02 henries carries 2 amps which is reduced to zero in 0.8 sec. (a) Find the initial energy stored in the magnetic field. (b) Find the average emf induced in the coil.

30.4   Challenge problem: From $U = LI^2/2$ and results in problem 30.2 show that the energy in the B-field is $U = \tfrac{1}{2}B^2/\mu_0$. This result is general and is the energy in magnetic fields including electromagnetic waves.

Mutual Inductance

When the current in some primary coil (primary=1) causes an induced current in a secondary coil (secondary=2), the induced emf in the secondary is given by

$$\varepsilon_2 = -M \frac{\Delta I_1}{\Delta t} \quad \text{or} \quad \varepsilon_2 = -M \frac{dI_1}{dt}$$

where coefficient M is called the *mutual inductance*.

30.5   A long solenoid carries a current I. Another coil with a larger diameter is coaxial with the solenoid, as in the figure below.

(a) Find the total magnetic flux through the N_1 turns in the large circular coil due to the current I in the solenoid.

(b) Show the mutual inductance M of the system is given by $M = \dfrac{\mu_0 N_1 N_2 A_2}{\ell_2}$

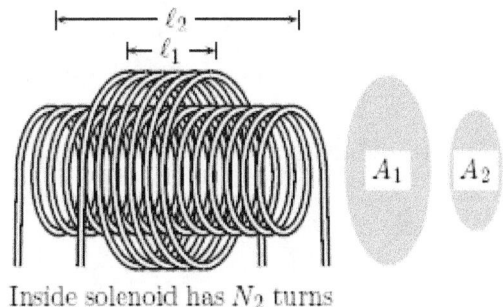

Inside solenoid has N_2 turns
Outside solenoid has N_1 turns

Answers

30.1   0.2 henries

30.3   0.04 J, 0.5 V

30.5   (a) $\varphi = I \left(\dfrac{\mu_0 N_2}{\ell_2} \right) N_1 A_2$

Problems

1. A 12 H inductor carries a current of 2.2 A. At what rate must the current be changed to produce a 81 V emf in the inductor?

2. Suppose that two insulated wires are wound onto a common cylindrical core of length 0.1 m and cross-sectional area 0.05 m^2. There are 100 turns in the first wire and 300 turns in the second wire. What is the mutual inductance of the two wires? If the current I_1 flowing in the first wire increases uniformly from 0 to 10 A in 0.1 s, what emf is generated in the second wire?

31

ALTERNATING CURRENTS

When current is made to oscillate by rapidly changing direction, the circuit can include capacitors and inductors.

Basic AC Circuit

An elementary AC circuit consists of an emf that varies sinusoidally

$$V = V_m \sin \omega t ,$$

and a resistor R, a capacitor C, and an inductor L, all connected in series in a loop. This is termed an RLC circuit. Other configurations are possible with these elements, but here we focus on this simple series circuit.

The current I and emf V vary sinusoidally with the same angular frequency ω, but their peaks and troughs need not match. When I and V match, they are said to be in *phase*.

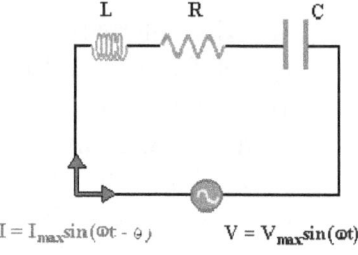

$I = I_{max} \sin(\omega t - \theta)$ $V = V_{max} \sin(\omega t)$

The amount of mismatch is measured by an angular quantity, θ, called the *phase angle*. (Recall that the relation between the frequency f and angular frequency is $\omega = 2\pi f$.)

Reactance

Reactance is a kind of generalized resistance due to circuit elements C or L opposing the flow of current:

resistance $\qquad\qquad X_R = R$

inductive reactance $\quad X_L = \omega L$

capacitive reactance $\;\; X_C = \dfrac{1}{\omega C}$

The effective resistance of the entire circuit is called *impedance* Z (note that it is not the sum of X's):

$$\text{impedance} \quad Z = \sqrt{(X_L - X_C)^2 + X_R^2}$$

An analog of Ohm's law holds between average V and I quantities called root-mean-squared (rms):

$$V_{rms} = I_{rms} Z$$

We often drop the subscripts "rms." The rms potential drop across each element is given by

$$V_R = I X_R, \quad V_L = I X_L, \quad V_C = I X_C$$

31.1 An RLC alternating current circuit has an effective voltage of 50 volts and is driven at an angular frequency of 1000 rad/sec. The inductor has a value of 0.9 H, the resistance 300 ohms, and the capacitance 2 μf
(a) Find the effective current through the-circuit.
(b) Find the potential drops across R, L, and C.

Phase Angle Diagram

The algebra relating impedance Z to the reactances is readily seen in the triangle figure. The phase angle θ is obtained from the diagram. The vector representing $V = IZ$ is said to be in the direction of the voltage and the segment IR is said to be in the direction of the current. When the angle is positive as shown, the "voltage leads the current." Equivalently, the "current lags the voltage." When the angle is negative, the "current leads the voltage."

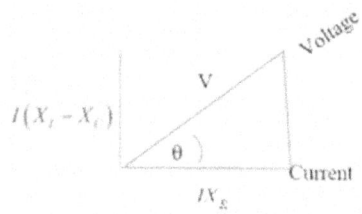

The power expended by an AC circuit is not simply $P=IV$. Rather,

$$P = I_{rms} V_{rms} \cos\theta$$

31.2 For the circuit of problem 31.1: (a) Find the phase angle between voltage and current. Indicate whether the voltage leas or lags the current. (b) Find the average power expended by the circuit. (c) Find the average power expended by the resistor alone.

31.3 Show from $V_{rms} = I_{rms}Z$ that the maximum current flow occurs when

$$\omega = 1/\sqrt{LC}.$$

This condition is called *resonance*. Calculate the resonant angular frequency for the system of problem 31.1.

Standard problems sometimes refer to maximum voltage or currents rather than their root-mean-square counterparts. The link between them is

$$V_{rms} = V_m / \sqrt{2},$$
$$I_{rms} = I_m / \sqrt{2}$$

Answers

31.1   (a) 0.1 A, (b) 30 V, 90 V, 50 V
31.2   53° voltage leads, 3 W, 3 W
31.3   745 rad/s

Problems

The next five questions pertain to the following circuit:

An RLC alternating current circuit has a rms voltage of 70 volts and is driven at an angular frequency of 100 rad/sec . The inductor has a value of 0.2 H, the resistance is 10 ohms, and the capacitance is 10^{-3} f.

1.   Calculate the impedance of the circuit.
      (a) 56 Ω, (b) 46 Ω, (e) 26.1 Ω, (d) 19 Ω, (e) 14.1Ω

2.   Find the rms current through the circuit.
      (a) 4.95 A, (b) 5.75 A, (c) 6.25 A, (d) 7.35 A, (e) 8.15 A

3.   Find the phase angle between voltage and current.
      (a) 60°, (b) 53°, (c) 37° (d) 30°, (e) 45°

4.   Calculate the voltage drop across the resistor alone.
      (a) 15.1 V, (b) 34 V, (c) 49.5 V, (d) 60 V, (e) 70 V

5.   What is the power dissipated by the circuit (in W)?
      (a) 100, (b) 145, (c) 200, (d) 245, (e) 300

32

THE QUANTUM LEVEL

This is a purely descriptive chapter. It is an overview of the quantum level and is intended to provide a framework for the more specific treatments to follow.

Prior units focused on theories of classical physics that were developed before the nineteen hundreds. The usefulness of these theories has not diminished with age. However, these theories do not describe the machinery of the universe at the fundamental levels of atoms, molecules, and elementary particles. This is the province of *modern physics* or *quantum physics*.

Quantum theory is the fundamental physical theory that evolved after centuries of scientific endeavor. Mechanics, electrodynamics, and thermodynamics are approximations to quantum theory–we say that they are "included" in quantum theory. Here we give a qualitative overview of quantum basics.

Light is a Wave

We've seen that light is a wave of electric and magnetic fields. One demonstration that shows light is a wave is the double slit experiment. Here light comes through two parallel slits. When the light reaches the screen, it forms a pattern of bars. Waves come from both slits and interfere at the screen; when two crests meet, they form a higher crest; a bright line. When a crest and trough meet, they cancel and produce a dark line. The results clearly demonstrate that light is a wave.

Light is a particle

In order to explain the *photoelectric effect,* Einstein had to interpret light as little lumps or particles of energy -- *photons.*

The photoelectric effect is seen in the following experiment: when blue light hits a particular metal, electrons fly out. But when red light hits this metal no electrons are ejected—even when the red light is intensely bright!

Einstein's explanation was that blue light (high frequency) has very energetic photons that can tear electrons out of the metal. Red light (low frequency) has less energetic photons that are unable to eject electrons. The energy of a light particle is proportional to the light's frequency,

$$E_{photon} = hf,$$

where h is a fundamental constant called *Plank's constant.*

32.1   *True or False*: For a particular metal, it is possible that green light will eject electrons, but ultraviolet waves will not.

We face the remarkable fact that light can have either wave or particle properties although these are very different things. Which of these properties is revealed depends on the situation.

Matter-Wave Duality

DeBroglie proposed that light should not be the only substance that exhibits both wave and particle properties. He said all waves have a particle aspect and that all particles have a wave aspect. In short, an object is both a particle and a wave and it reveals one property or the other depending upon the circumstances. Wave and particle properties were found to be linked such that wavelength λ is inversely proportional to momentum p

$$\lambda = h/p$$

This finding was verified by firing a beam of electrons through a crystal that served as a kind of "double slit." A pattern appeared on a screen showing characteristic wave interference.

32.2 *True or False*:
    (a) Light exhibits both wave and particle behaviors.
    (b) Protons exhibit both wave and particle behaviors

Bohr Atom

Bohr first calculated the energy levels of a hydrogen atom by imposing a non-classical requirement on the orbital angular momentum of the electron. The angular momentum (rp or rmv) was restricted to integer multiples of Planck's constant, h. The theory predicted that the atom could realize only discrete orbits with specific energy levels.

The restriction for orbits may be more readily seen as the result of an electron-wave fitting in standing patterns around the nucleus at just the appropriate distances. Waves may be imagined to look something like a string of sausages so that more sausages (wavelengths) can fit around larger orbits as seen in the diagram.

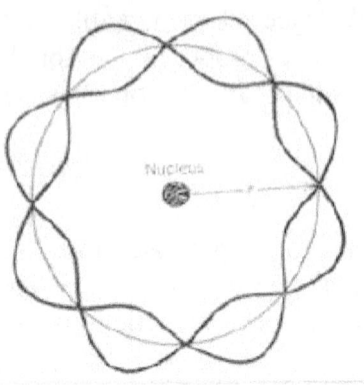

Schrödinger Equation

De Broglie waves are not adequate to treat any but the simplest atoms. Schrödinger constructed a more general wave equation that fully describes the waves for any system. Details of the Schrödinger equation are beyond the scope of this course. Howerver, a few generalities should be known in order to understand the periodic table and some atomic basics.

The Schrödinger Equation is the basic equation of non-relativistic quantum theory. There are two kinds of information given by the equation, eigenvalues and wavefunctions:

a)   eigenvalues
When applied to bound atoms and molecules, the Schrödinger equation predicts that the system has definite, discrete, energies:E_1. E_2, . No solutions exist for any energies "in-between." This is in sharp contrast with classical dynamics where there is a continuum of allowed energy values. The equation also determines that other quantities such as angular momentum have similar discrete values. However, solutions for unbound systems may also include quantities that are continuous (not discrete) such as the momentum of a free particle.

The above quantities that are given by the Schrödinger equation are examples of *eigenvalues;* the only values which the system can exhibit. In general, we look to the Schrödinger equation to provide the appropriate eigenvalues for a specific system.

32.3 *True or False*
    (a) Eigenvalues are numbers representing physical properties.
    (b) Eigenvalues are always discrete.
    (c) Eigenvalues represent the only measurable observables of a system.

b)   eigenfunctions

A *state* is characterized by a set of eigenvalues. For each state of the system, the Schrödinger equation gives a corresponding *eigenfunction* usually characterized by the Greek letter psi:

$$\Psi_1, \Psi_2, \Psi_3, \cdots .$$

Each eigenfunction is associated with specific set of observables like energy, angular momentum, and spin. The eigenfunctions belong to a more general category of waves called **wave functions** that are just linear combinations (weighted sums) of eigenfunctions. We have now to interpret the wavefunctions.

Born Interpretation

Physicist Max Born was the first to correctly interpret the wavefunction. Schrödinger and others initially thought Ψ was just a measure of the spread of a wavy particle. This belief had flaws that Born pointed out. Suppose that a single electron is shot through double slits. If the pattern that results has bright and dark fringes, then the electron has been divided into pieces.

However, electrons are never seen in parts. It must be that the particle itself is not actually an extended wavy lump. Rather, it is always a point particle when it is detected. The wave must be interpreted as the *probability* that the particle is in a particular position or state. (The larger the wave function at a point, the more likely is the particle to be at that location.)

32.4 *True or False:* Born interpreted eigenfunctions as extended wavy particles.

More About Wavefunctions

When Schrödinger introduced the wavefunction, he expected it to be something physical like the density of an extended particle. It turned out not to be a material substance. Rather, it is a ghostly probability wave that dictates the positions and outcomes of physical systems.

Ψ is not the probability itself. The probability of a state is found from the square of the wave function (much like the energy of an electromagnetic wave is given by the square of its fields).

32.5 *Question*: The diagram shows the eigenfunction for the location x of an electron in a particular molecule.

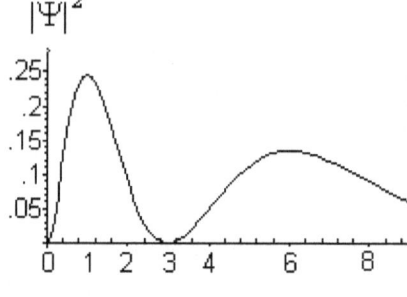

(a) Which labeled position is the most likely neighborhood for the electron?
(b) The electron is *never* at which of the labled points?

A related outcome in quantum physics is the *uncertainty principle* that says that a particle's position and velocity cannot both be measured with complete accuracy at the same time. One must sacrifice accuracy in one to increase accuracy in the other. Denoting the root-mean-squares of position and momentum as Δx and Δp respectively, this is expressed as $\Delta x \Delta p \geq \hbar/2$ where the symbol $\hbar$ ("h bar") is Planck's constant divided by 2π, $\hbar = h/2\pi$. A similar uncertainty exists between the lifetime and energy of a particle, $\Delta t \Delta E \geq \hbar$.

Answers

32.1 false
32.2 true, true
32.3 true, false, true
32.4 false
32.5(a) $x = 1$, (b) $x = 0$ and $x = 3$

33

PHOTONS

A wave is an extended object whereas a particle is treated as a point object. The fact that all particles have both wave and particle properties depending upon circumstances is challenging to the imagination. Nevertheless, this duality exists and we must accept it.

Photons

Einstein's expression for the energy E of a photon in a wave of frequency f is

$$E = hf$$

where $h = 6.63 \times 10^{-34}$ joule-sec is *Planck's constant.*

33.1 Calculate photon energy for light of wavelength 500 nm.

Photoelectric Effect

When metal is struck by a sufficiently energetic photon, some energy W is used to dislodge an electron and the remaining kinetic energy K is imparted to moving (and ejecting) the electron:

$$K = hf - W$$

where W is called the *work function* of the metal.

A unit of energy called the *electron volt*, eV, is very often used in modern physics. It is the energy an electron acquires when it is boosted through a 1-volt potential difference. The conversion factor is

$$1 \text{ eV} = 1.6 \times 10^{-19} \text{ J}$$

33.2   Light of wavelength 4300 Å strikes a sodium surface that has a work function of 2.28 eV. Calculate the (a) maximum energy of the emitted photoelectrons, and (b) the stopping potential (that is, the electrical potential that will stop an electron with energy K).

Photon Gas

A *blackbody* is an ideal substance that absorbs all radiant energy that falls on it. The body then emits radiant energy of all different frequencies. Planck treated blackbody radiation as a photon "gas" and derived the correct distribution of frequencies as they depend upon temperature T. We will not derive this expression here. Rather, we give two results that follow from this analysis; the *Stephan-Boltzmann law* and the *Wien displacement law*.

Stephan-Boltzmann law

The Stephan-Boltzmann law expresses the total power per unit area, S, being radiated from a blackbody:

$$S = \sigma T^4$$

$$\sigma = 5.67 \times 10^{-8} \text{ W/m}^2\text{-K}^4$$

33.3   A light bulb filament has an area of 1.27 X10-5 m2 and a temperature of 3200 K. What is the power expended by this filament?

Wien Displacement law

The Wien displacement law expresses the fact that as an object increases in temperature, the most abundant wavelength emitted, λ_{peak}, becomes progressively shorter in wavelength in inverse proportion to temperature,

$$\lambda_{peak} = \frac{2.90 \times 10^{-3}}{T} \, m$$

33.4   The most abundant wavelength in sunlight is about 500 nm. Estimate the temperature of the sun's surface.

Answers

33.1 $4.0 \times 10^{-19} \, J$
33.2   0.61 eV, 0.61 Volts
33.3   75.5 W
33.4   5800 K

Problems

1. The work function for copper is 4.7 eV. What is the incident frequency of radiation on copper in a photoelectric experiment that causes electrons to be emitted at a maximum speed of 2×10^6 m/s?

2. A star has a radius of 640,000 km. Spectral analysis shows that the temperature of this star is 5,800K. What is the total power it emits according to the Stefan-Boltzmann law? Treat the star as an ideal black body.

3. Blackbody radiation resulting from the Big Bang has a peak wavelength of about 1.07 mm. Use this to estimate the temperature of the microwave background throughout space.

34

MATTER WAVES

Matter waves are important in the history of quantum physics. Although the concept was superceeded by wave functions in modern quatum mechanics, matter waves are still useful for various applications and as a mental construct.

Matter-Wave Duality

De Broglie proposed that light is not the only thing that exhibits both wave and particle properties. He hypothesized that all waves have a particle aspect and that all particles have a wave aspect. In short, any object is both a particle and a wave and it shows one property or the other depending upon the circumstances. De Broglie expressed the wavelength λ of the matter-wave in terms of the momentum p of the particle.

The expression deduced by de Broglie applies to *all* matter including the most familiar particles; photons, electrons, protons, and neutrons:

$$\lambda = \frac{h}{p}$$

This conjecture was verified by shooting electrons through a crystal that served as kind of ultra fine diffraction grate. A pattern characteristic of wave interference appeared.

34.1 An electron microscope uses a beam of electrons boosted through a 10 V potential. Calculate the (nonrelativistic) momentum of these electrons and find their de Broglie wavelength. Compare this with visible light of wavelength 500 nm.

Application: Energy Levels

The following "particle-in-a-box" is a favorite model problem in quantum mechanics. The basic idea is that only an integer number n of half-wavelengths can form a standing wave in a one-dimensional "box." as illustrated in the diagram. Since the wavelength is associated with a momentum in $\lambda = h/p$, we find discreet values of momentum and energy for the particle.

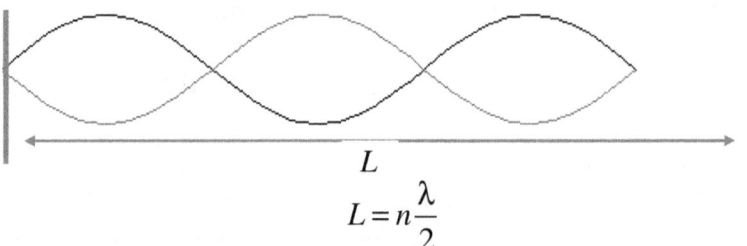

$$L$$

$$L = n\frac{\lambda}{2}$$

34.2 Derivation: A particle of mass m is confined in a one-dimensional box of length L such that its de Broglie wave must form a standing wave. Note that an integer number n of half wavelengths must fit in the box. Derive the allowed energy levels,

$$E_n = \frac{n^2 h^2}{8mL^2}.$$

34.3   Derivation: Derive the Bohr quantization rule $\ell = n\hbar$ for angular momentum by requiring that an integer number n of de Broglie wavelengths fit around the perimeter of a circular orbit. The symbol $\hbar$ is called "h-bar" and represents $h/2\pi$. It appears more often than h.

Characteristic Spectra

When an atom or molecule drops from a higher to a lower energy level, the excess energy is usually given off as a single photon. The photon energy hf is the difference between the levels,

$$hf = E_{high} - E_{low}$$

It follows that each atom or molecule emits only a particular set of radiation frequencies. This set of radiation frequencies is a virtual fingerprint of the substance.

34.5   A hydrogen atom has a ground state energy of -13.6 eV and a first excited state energy of -3.4 eV. Calculate the frequency of light emitted by a transition from the higher to the lower level.

Uncertainty Principle

The "spread" of a matter wave suggests that the location of the particle is somewhat indeterminate. If we take Δx and Δp to represent the amounts of uncertainty (usually the root mean square) of position and momentum, quantum theory shows that the more precise the position, the less precise the momentum. The relation is

$$\Delta x \Delta p \geq \frac{\hbar}{2}$$

34.6   Assume we can localize a particle to an uncertainty of 0.5nm. What will be the least uncertainty in the particle's momentum (in kg m/s)?

34 MATTER WAVES

Problems

1. Find the deBroglie wavelength of an electron after it is accelerated through a potential difference of 25 kV in a television set.

2. A hydrogen atom has first excited state energy of −3.4 eV and a second excited state of −1.5 eV. Calculate the wavelength of light emitted by a transition from the higher to the lower level.

3. Estimate the kinetic energy of an electron confined within an atom of size 0.53×10^{-10} m (radius of the hydrogen ground state) by using Heisenberg's uncertainty relation. Express your result in electron volts and compare it with the known ground state of −13.6 eV. Use the nonrelativistic form of kinetic energy $p^2/2m$.

HYDROGEN ATOM

The hydrogen atom is a standard model for atomic physics. It is the only atom for which the Schrödinger equation can be solved without approximations. It provides the basic framework for the periodic table of the elements.

Bohr Atom

The energy levels of a hydrogen atom can be derived from Bohr's requirement that angular momentum rmv is quantized so that $rmv = n\hbar$ where n is an integer. The result for the nth energy level is

$$E_n = -\frac{Rch}{n^2}$$

where the *Rydberg constant R* is given by

$$R = 1.097 \times 10^7 \text{ m}^{-1}$$

A more convenient form of this equation for numerical work is obtained by evaluating the constant product Rch:

$$E_n = -\frac{13.6}{n^2} \text{ eV}$$

The Bohr theory also derives the average orbital radius r_n for the nth energy level:

$$r_n = a_0 n^2$$

where the *Bohr radius* a_0 has the value

$$a_0 = 5.3 \times 10^{-11} \text{ m.}$$

Emitted Photon

The energy lost to an atom when it falls from a higher to a lower energy level is often emitted as a photon. The frequency of such a photon is then given in obvious notation by

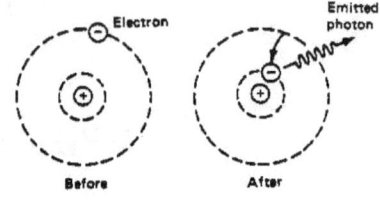

$$hf = E_{high} - E_{low}$$

35.1   For the hydrogen atom, $n=1$ is the *ground state* and $n=2$ is the *first excited state.*

(a) Calculate the energy and orbital radius of the ground state and the first excited state.

(b) Find the frequency of radiation emitted in a transition from the first excited state to the ground state

(c) Find the wavelength of the radiation in part b.

Bohr Theory

Bohr determined the hydrogen energy levels (and Rydberq constant) from the fundamental constants h, ε_0, e, and mass of the electron, m. He assumed that orbits are circles restricted to obey his quantization condition for angular momentum:

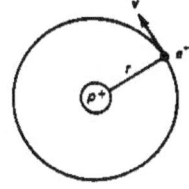

$$r\,mv = n\hbar$$

where $n = $ 1, 2, 3, $\cdots$.

35.2   Derivation: Derive the Bohr orbits and energy levels. Use the following steps:

1. Denote the orbital radius as r and the tangential velocity of the electron as v. Write Bohr's requirement for quantized angular momentum and solve this relation for v.

2. Write an equation to express the fact that Coulomb attraction provides the centripetal force to keep an electron in a circular orbit.

3. Substitute v from step 1 into the result of step 2. Solve for r. The result has the form $r = a_0 n^2$ where a_0 is the Bohr radius and n is an integer. Evaluate a_0..

4. Write an expression for conservation of energy for the Bohr atom. Replace v and r using your previous results to obtain the form $E = -C / n^2$ where C is a constant. Evaluate C in electron volts.

Denoting $k = 1/4\pi\varepsilon_0$ the results are:

$$r_n = \frac{\hbar^2}{ke^2 m} n^2$$

$$E_n = -\frac{k^2 e^4 m}{2\hbar^2 n^2}$$

Schrödinger's Solution

The simple Bohr picture of hydrogen gives the correct energy levels, but modern quantum mechanics gives much more information. In fact, all of chemistry is explained *in principle* by quantum theory. Here we will see that the Schrödinger solution for hydrogen gives us quantum numbers that reveal the basic plan behind the periodic table.

When we apply quantum theory to hydrogen, four eigenvalues (or associated quantum numbers) are needed to fully describe each state. These are

(1) the electron's energy level E_n characterized by the *principal quantum number n*, an integer corresponding to the levels in Bohr theory;

(2) the *orbital angular momentum L* of the electron characterized by an integer ℓ.

(3) the *azumuthal angular momentum L_z* is the z-component of *L*. It is characterized by an integer quantum number *m* that can be positive or negative.

(4) the intrinsic angular momentum of the electron spinning on its axis called the *spin*, s_z

The eigenvalues (or the corresponding quantum numbers) are related to each other. For a given energy, only a limited amount of angular momentum is possible and this restricts the values for ℓ and m. The rules restricting the quantum numbers are given below:

- n principal quantum number is unrestricted
$$n = 1, 2, 3, ...$$
- ℓ orbital angular momentum quantum number is limited to a value below the value of n,
$$\ell = 0, 1, 2, ..,n\text{-}1$$
- m the azumuthal quantum number can be negative or positive, but may not exceed the magnitude of the total orbital angular momentum,
$$m = 0, \pm 1, \pm 2,.., \pm \ell$$
- s_z spin quantum number can only have one of two values,
$$s_z = \pm \tfrac{1}{2}$$

35.3 Use the rules to find how many hydrogen electron states are possible when
(a) $n = 1$, and (b) $n = 2$.

Many-Electron Atoms

Pauli's exclusion principle asserts that *no two electrons can occupy the same quantum state*. For atoms with many electrons, this implies that only 2 electrons can be in the $n=1$ ground state. Similarly, there are 8 different states with the next higher energy level ($n=2$). We say that a maximum of 8 electrons can exist in this energy level. At the $n=3$ level, the simple counting rules for hydrogen's electron must be modified to account for the compli-

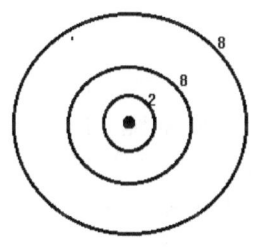

cating effects of additional electrons in many-electron atoms. The result for $n=3$ is that 8 states are allowed.

2 electrons are allowed in the lowest energy level, $n=1$.

8 electrons are allowed in the second energy level, $n=2$.

8 electrons are allowed in the third energy level, $n=3$.

We have treated each n level or *shell* as if all the electrons therein have exactly the same energy. This is not strictly true for many-electron atoms. Very small differences in energy occur for the different ℓ values due to various interactions between electrons. For example, when these complications are included for $n = 2$, the $\ell = 0$ state is found to be slightly lower in energy than the $\ell = 1$ state. Electrons with different values of ℓ within the same shell are therefore called *subshells*.

The number of electrons in each subshell is the product of the number of m values ($2\ell + 1$) and the number of spin states, 2:

maximum number of electrons in a subshell $= 2(2\ell + 1)$

Some historic names for the ℓ values are

$\ell = 0$	s wave
$\ell = 1$	p wave
$\ell = 2$	d wave
$\ell = 3$	f wave

and the subshells are denoted by the combination "$n\ell$" so that the $n=1$, $\ell = 0$ subshell is written 1s and the $n=2$, $\ell = 1$ subshell is written 2p. (This is called "spectral notation.")

35.4   Write the spectral notation for the $n = 4$, $\ell = 3$ subshell.

35.5   Write the spectral notation for sodium (eleven electrons).

Periodic Table

The rules we've established for three energy levels can be used to construct a simplified periodic table. The elements in each row or *period* have their outer electrons occupying the same shell or *n*-quantum number.

Each column or *group* has similar chemical properties because the elements in the group have the same number of outer electrons (called *valence* electrons). These are the electrons that encounter the surrounding world and account for most chemical interactions.

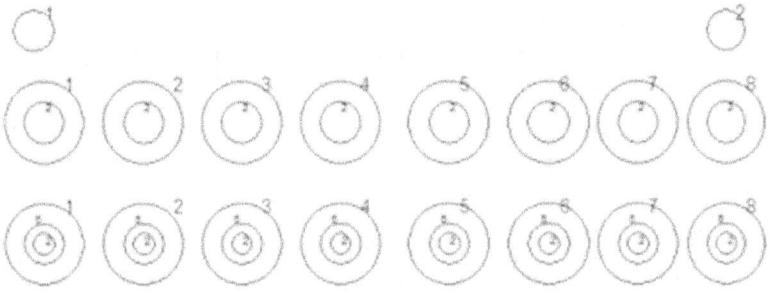

This is a starkly abridged periodic table, but it displays the underlining fundamental themes. Notice that the inner orbits get smaller as *n* increases because more protons in the nucleus exert a greater electric force on the inner electrons. Conversely, the outer electrons are more loosely bound because they are further from the nucleus and they are shielded from the protons by the inner electrons. Shielding increases for outer electrons toward the right of the table so these atoms are more disposed to surrendering their valence electrons.

35.6 The diagram depicts the orbital electrons of Lithium. Without reference to the periodic table, make similar sketches for the elements just to the right and just below Lithium on the periodic table.

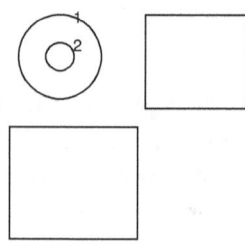

35.7 Choose the correct statement. Elements in the same *group* of the periodic table have
(a) different chemical properties but a similar number of energy shells
(b) different chemical properties but a similar numbers of outer electrons
(c) similar chemical properties and a similar number of outer electrons
(d) similar chemical properties and a similar number of energy shells

35.8 Choose the correct statement. The ground state configuration of a particular element is $1s^2$, $2s^2$, $2p^6$ $3s^2$, $3p^1$. Let ℓ be the angular momentum quantum number. The outermost energy level contains
(a) 2 $\ell=0$ electrons and 3 $\ell=1$ electrons,
(b) 2 $\ell=0$ electrons and one $\ell=1$ electron,
(c) 5 $\ell=0$ electrons and 3 $\ell=1$ electrons ,
(d) 2 $\ell=1$ electrons and 3 $\ell=2$ electrons ,
(e) 8 $\ell=0$ electrons and 8 $\ell=1$ electrons

Answers

35.1 (a) ground -13.6 eV, 5.3×10^{-11} m, excited -3.4 eV, 2.1×10^{-10} m,
      (b) 2.46×10^{15} Hz (c) 122 nm
35.3 2, 8
35.4 4d
35.5 $1s^2 2s^2 2p^6 3s^1$
35.7 c
35.8 b

Problems

1. Use the Bohr derivation to find expressions for the radii and energy levels of an ionized atom having Z protons in its nucleus and only one orbiting electron. Show that the radii vary inversely with Z and binding energy levels are proportional to Z^2.

2. A hydrogen atom is in a3D state. The atom decays to a 2D state by emitting a photon. Find the photon energy in eV.

3. An exotic atom has a pion orbiting the nucleus rather than an electron. The pion has no spin, but otherwise the Schrödinger equation gives the same restrictions between n, ℓ, and m as for the hydrogen atom. Determine the number of possible states for $n=1$ and $n=2$

PHYSICAL CONSTANTS

Constant	Symbol	Value
acceleration due to gravity	g	9.8 m s^{-2}
atomic mass unit	amu, m_u or u	1.66×10^{-27} kg
Avogadro's Number	N	6.022×10^{23} mol^{-1}
Bohr radius	a_0	0.529×10^{-10} m
Boltzmann constant	k	1.38×10^{-23} J K^{-1}
electron charge to mass ratio	$-e/m_e$	-1.7588×10^{11} C kg^{-1}
electron classical radius	r_e	2.818×10^{-15} m
electron mass energy (J)	$m_e c^2$	8.187×10^{-14} J
electron mass energy (MeV)	$m_e c^2$	0.511 MeV
electron rest mass	m_e	9.109×10^{-31} kg
Faraday constant	F	9.649×10^4 C mol^{-1}
fine-structure constant	α	7.297×10^{-3}
gas constant	R	8.314 J mol^{-1} K^{-1}
gravitational constant	G	6.67×10^{-11} Nm^{2}kg^{-2}
neutron mass energy (J)	$m_n c^2$	1.505×10^{-10} J
neutron mass energy (MeV)	$m_n c^2$	939.565 MeV
neutron rest mass	m_n	1.675×10^{-27} kg
permeability of a vacuum	μ_0	$4\pi \times 10^{-7}$ N A^{-2}
permittivity of a vacuum	ε_0	8.854×10^{-12} F m^{-1}
Planck constant	h	6.626×10^{-34} J s
proton mass energy (J)	$m_p c^2$	1.503×10^{-10} J
proton mass energy (MeV)	$m_p c^2$	938.272 MeV
proton rest mass	m_p	1.6726×10^{-27} kg
proton-electron mass ratio	m_p/m_e	1836.15
Rydberg constant	r_∞	1.0974×10^7 m^{-1}
speed of light in vacuum	C	2.9979×10^8 m/s

APPROXIMATE CONVERSIONS TO SI UNITS

SYMBOL	GIVEN	MULTIPLY BY	TO FIND	SYMBOL
LENGTH				
in	inches	25.4	millimeters	mm
ft	feet	0.305	meters	m
yd	yards	0.914	meters	m
mi	miles	1.61	kilometers	km
AREA				
in^2	square inches	645.2	square mm	mm^2
ft^2	square feet	0.093	square m	m^2
mi^2	square miles	2.59	square km	km^2
VOLUME				
fl oz	fluid ounces	29.57	milliliters	mL
gal	gallons	3.785	liters	L
ft^3	cubic feet	0.028	cubic m	m^3
MASS				
oz	ounces	28.35	grams	g
lb	pounds	0.454	kilograms	kg
TEMPERATURE				
^{o}F	Fahrenheit	5 (F-32)/9	Celsius	^{o}C
FORCE and PRESSURE or STRESS				
lb	pound	4.45	newtons	N
lb/in^2	pound/sq inch	6.89	kilopascals	kPa
1 atm	atmosphere	1.013×10^5	Pascals	Pa

Index